SPECIAL RELATIVITY:
A Modern Introduction

Hans C. Ohanian

Physics Curriculum & Instruction

Printed in the United States of America

First Edition

Library of Congress Control Number: 2001093988

ISBN 0-9713134-1-5

Physics Curriculum & Instruction, Inc.
22585 Woodhill Drive
Lakeville, MN 55044 USA
www.physicscurriculum.com

To V. T. P.

Contents

PREFACE

When Einstein published his theory of Special Relativity in 1905, its profound significance was quickly recognized by the best and most influential theoretical physicists of the time. Planck immediately became the leading supporter of the theory, and he later secured for Einstein the prestigious appointment in Berlin. However, experimental support for the theory--other than the Michelson-Morley experiment and repetitions thereof--was much slower to develop. Experiments by Kaufmann in 1905 on the nonlinear increase of momentum with velocity ("increase of mass with velocity") disagreed with the prediction of Special Relativity. Some years later, more precise experiments by Bucherer and others seemed to show agreement with the relativistic formula, and Bucherer claimed, "By means of careful experiments, I have elevated the validity of the principle of relativity beyond any doubt." These experiments led to wide acceptance of the theory. With the wisdom of hindsight, we know that these claims were premature. A later reanalsyis of the experimental data was to show that all the early experiments were afflicted with much larger experimental uncertainties than claimed, so these experiments merely established a qualitative increase of mass with velocity but gave no adequate quantitative confirmation of the relativistic formula.

By now, nearly a hundred years have elapsed since the inception of Special Relativity, and all aspects of the theory have been tested and retested by a long succession of experiments, which give it firm support. The only discrepancies between the theory and experiment or observation involve gravitational effects, such as the gravitational redshift, the perihelion precession of planets, and the deflection of light by the Sun, which lie outside of the realm of applicability of Special Relativity and require General Relativity for their explanation.

Not only is Special Relativity an essential part of the physics of today, but it has a variety of practical applications, such as the design of high-voltage TV and X-ray tubes, the design of particle accelerators (some of which are used for medical therapy and the industrial production of radioisotopes), and the Global Positioning System and Very Long Baseline Interferometry (GPS and VLBI, in both of which relativistic corrections to clock rates and synchronization play a role).

It is therefore fitting that Special Relativity should be an integral part of the undergraduate curriculum, not only for physicists but also for engineers. In Spain, in recognition of the central role of relativity in modern physics, it is included as a required part of the physics curriculum for high school students seeking to graduate with a *bachillerato* in science.

The present book is an introduction to relativity for first- or second-year university students. It can serve as the textbook for a separate short course or as a supplement to an introductory general physics course or an electromagnetism course.

In essence, relativity is a theory of the geometry of spacetime, and I therefore emphasize the geometrical aspects of relativity throughout the book. To encourage geometrical thinking, I use a profusion of spacetime diagrams for graphical explanations and derivations of the Lorentz transformation, the time-dilation and length-contraction paradoxes, etc. I also emphasize the importance of experiments in the study of relativity, and I include fairly complete, but not exhaustive, summaries of relevant experiments up to 2000.

Chapter 1 is a review of relativity in Newtonian physics and of the Michelson-Morley experiment and its modern versions. Chapter 2 introduces Einstein's postulates for relativity and derives the Lorentz transformation by means of graphical methods. Chapter 3 deals with the time dilation, the length contraction, and the associated paradoxes. Chapter 4 formulates the relativistic laws of dynamics and the relativistic expressions for momentum and energy. Chapter 5 applies these laws to collision problems, and it includes a detailed treatment of the elastic collision in one dimension (oddly, this problem is hardly ever treated in relativity textbooks). Chapter 6 examines the implications of relativity for electricity and magnetism; this chapter relies on a simplified version of the argument of Frisch and Wilets to derive magnetic forces and fields from electric forces and fields. Finally, Chapter 7 contains a brief introduction to General Relativity.

It is a pleasure to thank several colleagues who read the manuscript of this book and gave me the benefit of their criticism. I am especially grateful to Professor Antonio Bernalte Miralles, UNED Madrid, who not only gave me many valuable suggestions for improvements of the content and organization of the chapters and sent me copious extracts from books I did not know, but also invited me to present lectures on geometrical methods at the UNED Summer Schools on Relativity in Plasencia (1998) and Avila (2000), which encouraged me to develop and refine the graphical methods of Chapters 2 and 3. I am grateful to Professor Francisco Ares Peña, Universidade de Santiago de Compostela, for bringing to my attention several papers on the connection between relativity and electromagnetism, among them the paper by Frisch and Wilets on which Chapter 6 is based. I thank Professors Nathaniel P. Longley, Swarthmore College, and Robert L. Lysak, University of Minnesota, for their detailed reviews of drafts of the book. And last, but not least, I thank my friend and editor Stephen Mosberg for his enthusiastic support of the publication of this book and for his perceptive comments on an early draft.

H.C.O. Salamanca, July 2000

1

Relativity in Newtonian Mechanics and the Michelson-Morley Experiment

NEWTON, FORGIVE ME . . . THE CONCEPTS THAT YOU CREATED ARE EVEN TODAY STILL GUIDING OUR THINKING IN PHYSICS, ALTHOUGH WE NOW KNOW THAT THEY HAVE TO BE REPLACED BY OTHERS, MORE REMOTE FROM THE REALM OF DIRECT EXPERIENCE.

Albert Einstein, *Autobiographical Notes*

The theory of relativity has its roots in the empirical observation that a uniform translational motion of a physical system has no effect on the phenomena that occur in this system. This was first recognized by Galileo Galilei in the 17th century, who in his *Dialogue of the Two Principal Systems of the World* wrote:

> Shut yourself up with some friend in the largest room below decks on some large ship, and there procure gnats, flies, and such other small winged creatures. Also get a great tub full of water and within it put certain fishes; let also a certain bottle be hung up, which drop by drop lets forth its water into another narrow-necked bottle placed underneath. Then, the ship lying still, observe how those small winged animals fly with like velocity towards all parts of the room; how the fishes swim indifferently towards all sides; and how the distilling drops all fall into the bottle placed underneath. And casting anything towards your friend, you need not throw it with more force one way than another, provided the distances be equal; and jumping broad, you will reach as far one way as another. Having observed all these particulars (though no man doubts that, so long as the vessel stands still, they ought to take place in this manner), make the ship move with what velocity you please, so long as the motion is uniform and not fluctuating this way and that. You shall not be able to discern the least alteration in all the forenamed effects, nor can you gather by any of them whether the ship moves or stands still. [1]

Thus, the uniform motion of the ship is relative. Such motion can be detected only by measuring the position of the ship in relation to some external landmark; it cannot be detected by any mechanical experiment within the ship.

When Newton formulated his laws of motion, later in the 17th century, he deduced the relativity of uniform motion from these laws and stated this as Corollary V in his *Principia*:

> The motions of bodies included in a given space are the same among themselves, whether that space is at rest, or moves uniformly forwards in a right line . . . A clear proof of this we have from the experiment of a ship; where all motions happen after the same manner, whether the ship is at rest, or is carried uniformly forwards in a right line. [2]

Both Galileo and Newton took a ship as an example, because ships were the only vehicles of their days capable of a fairly uniform motion. Today, we can draw on our wide experience with the uniform motion of trains, aircraft, or automobiles, and our experience confirms that such motion cannot be detected by phenomena within the vehicle. Likewise, the translational motion of the Earth in its orbit around the Sun cannot be detected by mechanical experiments within the laboratories on the Earth--we feel nothing of this motion, and it produces no observable effects on any phenomena on the Earth.

However, after the discovery of the wave properties of light, the physicists of the 19th century became convinced that uniform motion should be detectable by experiments on the propagation of light. In essence, they argued that it should be possible to detect the translational motion of the Earth by measuring this motion relative to light waves propagating in laboratories on the Earth. Several experiments of this kind were attempted, culminating in the famous Michelson-Morley experiment, originally performed in 1881 and repeated many times since then. But all these experiments gave null results--the translational motion of the Earth had no effect whatsoever on the observed propagation of light waves in laboratories on the Earth. This indicated some serious defect in the foundations of Newtonian physics. Nineteenth-century physicists proposed several explanations of the null results of their experiments, but none of these explanations proved satisfactory. That set the stage for Einstein's revolutionary ideas about space and time and for the replacement of Newtonian physics by a new, relativistic physics.

We will begin this chapter with the definition of the inertial reference frames used for the measurement of space and time. Under the assumption that Newtonian physics is valid, we will see how positions, times, and velocities in different reference frames are related. Such a review of Newtonian physics will provide some helpful background for our treatment of relativistic physics in later chapters.

We will then discuss why Newtonian physics predicts that the propagation of light waves can be used to detect uniform translational motion, and we will discuss the Michelson-Morley experiment and related experiments, as well as 19th-century attempts at explaining the null results of these experiments.

1.1 INERTIAL REFERENCE FRAMES

Like a detective investigating a crime, a physicist investigating a phenomenon must begin by asking where did it happen and when did it happen. Phenomena happen at points of space and at instants of time. For instance, a collision between two pointlike particles is a very simple phenomenon that occurs at one point of space at one instant of time. A more complicated phenomenon, such as motion of a particle in a circular or an elliptical orbit around a gravitating mass, involves many points of space and time. To describe this phenomenon, the physicist needs to examine the behavior of the particle at different positions and times along its orbit. A happening at one point of space at one instant of time is called an **event**. Any phenomenon can be described as a succession of such individual events.

To determine the location of events in space and time, physicists use reference frames. A **reference frame** is a three-dimensional lattice of measuring rods and clocks that permits us to assign space and time coordinates to an event. Figure 1.1 shows a reference frame consisting of a rectangular lattice of measuring rods, with clocks at the intersections of the measuring rods. One point within this lattice is

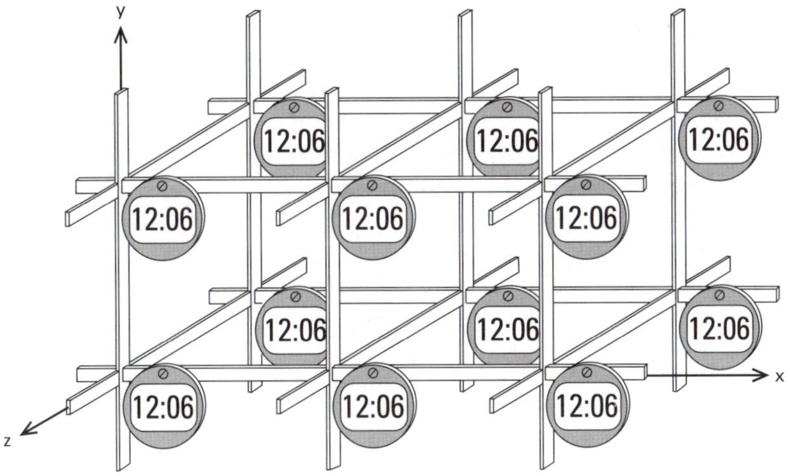

Fig. 1.1 A reference frame consisting of a rectangular lattice of measuring rods and synchronized clocks. The portion of the lattice shown here is to be continued indefinitely in all directions, so that it spans all of space.

taken as origin, and the x,y,z distances from the origin to an event give us the rectangular x,y,z coordinates of the event. The time coordinate t of the event is obtained by reading the time registered by the clock at the event. Instead of the rectangular coordinates x,y,z, we could use cylindrical coordinates, or spherical coordinates, or some other coordinates; but for the purposes of relativity, rectangular coordinates are the most convenient.

We will assume that all our reference frames are **inertial**. This means that the reference frame must meet two conditions: Newton's First Law must be valid, and the clocks of the reference frame must be synchronized. Newton's First Law governs the motion of free particles, that is, particles free of external forces. This law states that any free particle at rest remains at rest, and any free particle in motion continues to move along a straight line with constant velocity. Accordingly, the inertial reference frame must have a uniform, unaccelerated motion relative to free particles. To find an inertial reference frame, we can use a simple trial-and-error procedure: we place a few free particles, initially at rest, along the axes of the reference frame; if these particles remain at rest, the reference frame passes the test; if not, it fails, and it must be replaced and retested. The synchronization of the clocks of the reference frame is a more delicate matter and requires some discussion of the properties of time in Newtonian physics and the effects of motion on the behavior of clocks.

In Newtonian physics, both length and time are **absolute**. This means that the lengths of measuring rods and the rates of clocks do not depend on their position or on their motion. For instance, consider two identically constructed clocks, which are initially synchronized while they are together at the same location. The absolute character of time implies that if we transport these clocks along separate trajectories and finally bring them back together, they will still be synchronized. This assumes that we employ suitably constructed clocks that are insensitive to the accelerations to which they are subjected during their transport (a good mechanical or electronic wristwatch is quite insensitive to acceleration--even if you shake your arm violently, the rate of the watch is unaffected).

Newton and his contemporaries deemed this absolute character of time self-evident, and they did not consider it necessary to perform any direct experimental tests. However, during the 18th and 19th centuries, precise tests of the absolute character of time were performed indirectly, in the course of determinations of geographical longitude. The standard method for the determination of longitude at a given location relies on the comparison of local solar time and Greenwich time, that is, the local solar time kept by clocks at the Greenwich observatory near London, on the zero meridian (you can listen to telephone announcements of Greenwich time, or "Coordinated Universal Time," by dialing 303 499 7111). The longitude is determined from the difference between these times, 360° of longitude for

24 hours of difference, or 15^0 for 1 hour. For such longitude determinations, ships carried chronometers synchronized to Greenwich time. These chronometers were routinely compared with clocks at harbors at the beginning and the end of voyages, to verify that they had maintained their Greenwich synchronization. The best 18th- and 19th-century chronometers maintained synchronization to within a random error of a few seconds after voyages of several months, that is, a random error of a few parts in 10^7. Surveying ships sometimes carried as many as 40 chronometers, so that the random error could be reduced by averaging their readings. This reduced the synchronization error to somewhat below one part in 10^7 and verified the absolute character of time to within this same level of precision.

More precise tests of the absolute character of time became possible with the development of portable atomic clocks around 1970. Portable rubidium and cesium atomic clocks were carried in aircraft back and forth between laboratories on the Earth, and they were installed on satellites orbiting the Earth, where they are used for the operation of the Global Positioning System (GPS). These atomic clocks have detected deviations from the absolute character of time at a level of a few parts in 10^{10}. These deviations from Newtonian physics are relativistic effects, and we will examine them in later chapters. But in this first chapter we will ignore these very small deviations, and we will assume that time is absolute, in accordance with Newtonian physics.

If time is absolute, we can achieve synchronization of the clocks in a reference frame by means of **clock transport**. To synchronize the clocks at different locations in the reference frame, we simply transport a portable clock successively to each of these different locations. We initially synchronize the transported clock with the first clock at the first location, and we then synchronize each successive clock with the traveling clock when it passes by its location. This transfers the synchronization from the first clock to all the other clocks. Furthermore, we can synchronize all the clocks of the given reference frame with all the clocks of any other reference frame--we merely let the transported clock pass through the second reference frame and again synchronize each clock with it. Thus, the absolute character of time permits us to synchronize every clock in every reference frame with every clock in every other reference frame.

It has sometimes been claimed that such synchronization by clock transport is merely a convention, no better or worse than some other convention for synchronization. But this is false. Synchronization by clock transport plays a preferential role in Newtonian physics. It is the only synchronization consistent with the absolute character of time. Furthermore, it is the only synchronization consistent with Newton's Second Law, $m\mathbf{a} = \mathbf{F}$, since this law incorporates the absolute character of time. We can recognize this from two results that can be deduced from the Second Law: any isolated oscillating or rotating periodic system (for example, two masses joined by a spring) has the same period when at rest and when in uniform translational motion; and if the periodic system consists of a rigid body rotating about an axis through its center of mass (for example, the balance wheel of a watch, with an ideal massless torsional spring), then it has the same period even when in nonuniform translational motion. Thus, Newton's Second Law implies that a clock incorporating such a periodic system remains in synchronization with the clocks of the inertial reference frame when it is transported.

Of course, for practical reasons we might occasionally replace clock transport by some other mechanical synchronization procedure. For instance, we might use interlocking gears, or a drive shaft, or a chain drive to link two clocks at separate locations and thereby synchronize them; or we might use sound waves in air or some other elastic medium to send signals from one clock to another, provided we know the speed of the sound waves so that we can take into account the propagation delay. But all such mechanical procedures are ultimately based on Newton's Second Law, and they give synchronizations identical to that given by clock transport. Mechanical procedures are very useful for the synchronization of clocks, or clock-like machinery, over short distances. For example, automobile engines use a "timing chain" to link the crankshaft and the camshaft, to synchronize the rotation of these shafts. But for synchronization over long distances, such mechanical procedures are impractical, and clock transport is a better choice.

The importance of the correct synchronization of the clocks is often overlooked in discussions of inertial reference frames. It is crucial to recognize that deviations from correct synchronization are just as disastrous for an inertial reference frame as deviations from uniform motion--both of these kinds of deviations lead to violations of Newton's Second Law. This does not mean that we cannot do physics in a reference frame with nonuniform, accelerated motion or in a reference frame with an incorrect synchronization. In an accelerated reference frame we can write a modified Second Law, which includes pseudoforces, such as the centrifugal and Coriolis pseudoforces found in a rotationally accelerated reference frame. These pseudoforces compensate for the violations of the Second Law that arise from the acceleration. Likewise, in an incorrectly sychronized reference frame, we can write a modified Second Law with pseudoforces that compensate for the violations. But the presence of such pseudoforces reveals that the reference frame is *not inertial*. In an inertial reference frame, Newton's Second Law must be valid without any such modifications, and therefore any acceleration of the reference frame or incorrect synchronization of the clocks is forbidden.

Note that a reference frame attached to the surface of the Earth is not an inertial reference frame, even if we adopt Greenwich time for all points on the Earth. Because this reference frame participates in the rotational motion of the Earth about its axis, it has an accelerated motion, with a centripetal acceleration directed toward the axis.

The rotational motion induces pseuodoforces, which produce effects that are readily observable in laboratories on the Earth, such as the precession of the plane of oscillation of a Foucault pendulum and the precession and north-south alignment of the spin axis of a gyrocompass. However, a reference frame that participates in the translational motion of the Earth but not in its rotational motion is an inertial reference frame. This is true even though the orbital motion of the Earth around the Sun consists of translational motion tangential to the orbit and a simultaneous centripetal acceleration toward the Sun. This centripetal acceleration is a motion of free fall in the gravitational field of the Sun, and the observable effects of this centripetal acceleration are canceled by the effects of the gravitational field. We will deal with this in more detail in Chapter 7, when we discuss the Principle of Equivalence of gravitation and acceleration. Throughout this book, whenever we speak of the inertial reference frame of the Earth, we will always assume that this reference frame has the translational motion of the Earth, but not the rotational motion, and we will call this the reference frame **centered** on the Earth, to distinguish it from a reference frame attached to the surface of the Earth and rotating with it.

1.2 SPACETIME DIAGRAMS AND WORLDLINES

In a given inertial reference frame, the motion of a particle can be represented graphically in a **spacetime diagram**, that is, a plot of the position coordinates x,y,z as a function of time. Each point in the spacetime diagram represents the position of the particle at one instant of time, that is, it represents an event. The general spacetime diagram is four-dimensional, with three space axes and one time axis, and it is impossible to draw it on a page. The best we can do is to draw three of the four axes, say, two space coordinates and the time coordinate. Figure 1.2 is a simple example of the graphical representation of the motion of a particle in such a spacetime diagram. Initially this particle is at rest at $x = x_0$ (straight vertical line for $t < 0$); then it accelerates in the x direction (curved line); and then it continues to move with constant velocity in the x direction (straight slanted line). Of course, Fig. 1.2 is merely the familiar plot of x vs. t often used in kinematics, but in relativistic physics it is customary to plot t vertically and x horizontally. The curve representing the motion of a particle in the spacetime diagram is called the **worldline** of the particle. The worldline is a graphical summary of all the information about the motion of the particle--it is a graphical history of the motion of the particle. Note that because of the vertical orientation of the time axis, a vertical worldline represents a particle at rest, and a worldline of small slope (nearly horizontal) represents a particle moving at very high speed.

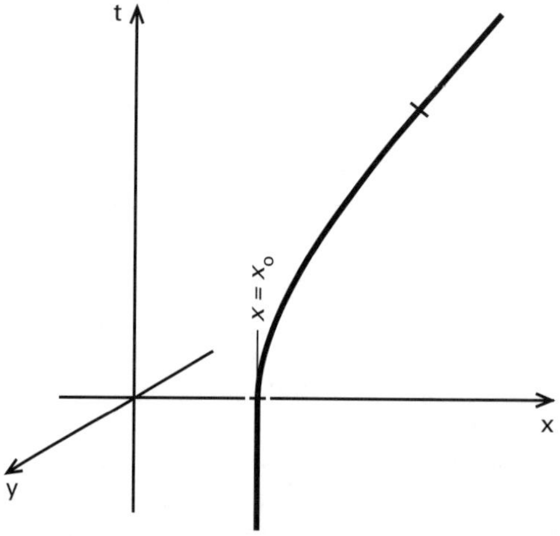

1-2

Fig. 1.2 Spacetime diagram showing the worldline of a particle that is initially at rest, then accelerates briefly, and then continues to move with constant velocity.

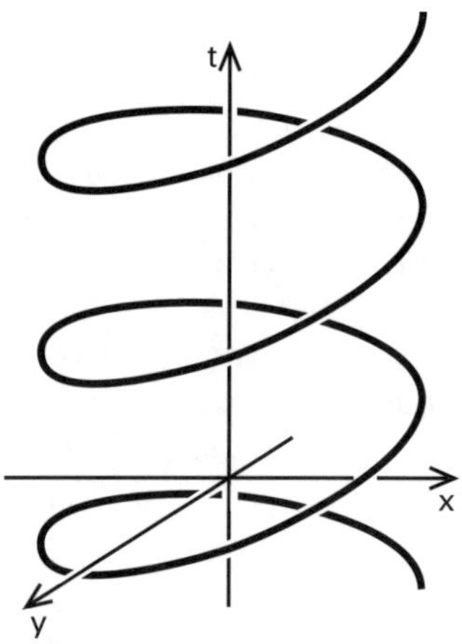

1-3

Fig. 1.3 Spacetime diagram showing the worldline of a particle moving along a circular orbit in the x-y plane with constant speed.

Figure 1.3 is another example of a spacetime diagram. This diagram describes the motion of a particle moving along a circular orbit with constant speed. In the x–y plane this orbit is a circle, but in spacetime the corresponding worldline is a helix.

■ **EXAMPLE 1.** Two particles of equal masses collide elastically. Initially, one of the particles is at rest at $x = x_0$ on the x axis, and the other is moving toward the right along the x axis at speed v. Sketch the worldlines of the particles before and after this collision.

SOLUTION. This is a well-known example of a one-dimensional elastic collision, for which the laws of conservation of energy and momentum completely determine the final speeds in terms of the initial speeds. After the collision, the particle 1 that was at rest will be moving at speed v, and the particle 2 that was moving will be at rest. Figure 1.4 shows the worldlines before the collision ($t < 0$) and the worldlines after the collision ($t > 0$). Before the collision, the particle 1 resting at $x = x_0$ has a vertical worldline, and the moving particle 2 has a worldline slanting toward the right. After the collision, particle 1 has a worldline slanting toward the right, with the same slope, and particle 2 has a vertical worldline.

According to Newton's First Law, any free particle moves along a straight line in space, and it has a constant velocity. In the spacetime diagram we can express both of these conditions geometrically by

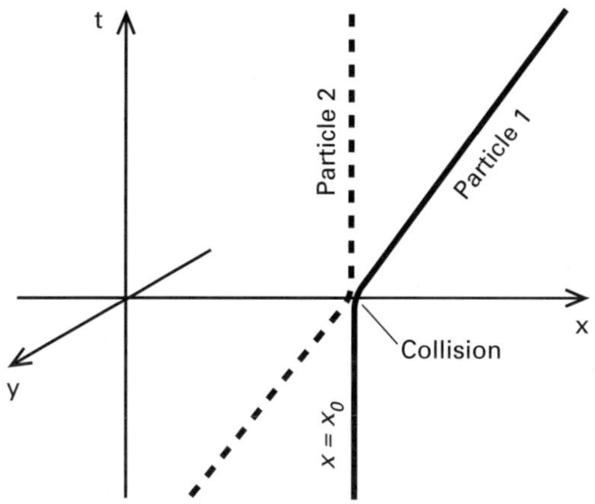

Fig. 1.4 Worldlines of two particles 1 and 2. The particles collide at $x = x_0$ at $t = 0$.

saying that the free particle has a straight worldline. For such a straight worldline, the plot of each space or time coordinate against any other space or time coordinate is a straight line, which confirms that the motion proceeds along a straight line in space and also confirms that it proceeds with constant velocity components v_x, v_y, and v_z, that is, constant slopes in the plots of x, y, and z vs. time.

1.3 THE GALILEAN TRANSFORMATION

Inertial reference frames can differ by the orientation of their axes, and they can differ by a displacement of their origins. A change of orientation is a rotation (by a constant angle), and a displacement of the origin is a translation (by a constant distance). Rotations and translations are examples of transformations of reference frames, but we will ignore these simple transformations because they are not directly relevant to relativity.

For the purposes of relativity, the interesting difference between two inertial reference frames is a difference in their translational motion. Figure 1.5 shows two reference frames x,y,z and x',y',z' with different translational motions. These two reference frames have the same orientation, but the displacement between their origins increases steadily and uniformly with time. The restriction to uniform relative motion is required by the inertial character of the reference frames; any nonuniform motion would imply that at least one of the reference frames is not inertial. (If a free particle remains at rest in, say, the first reference frame, then this free particle has a uniform motion relative to the second reference frame if and only if the relative motion of the reference frames is uniform.) For convenience, the direction of the relative velocity in Fig. 1.5 has been chosen along the x axis of each reference frame. The second reference frame travels toward the right along the x axis of the first, and the first travels toward the left along the x' axis of the second. If the speed of each

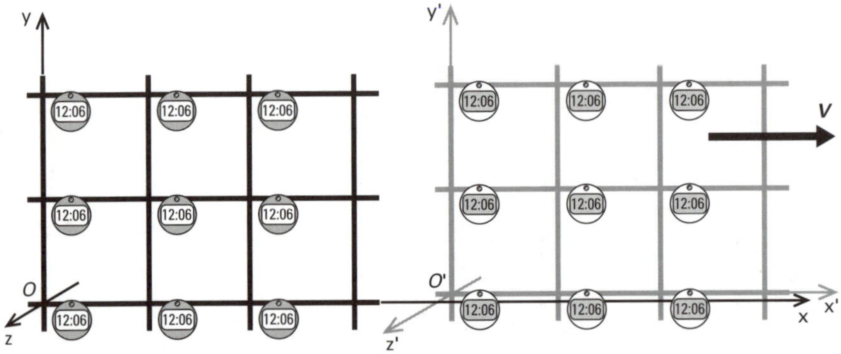

Fig. 1.5 Two reference frames with a relative velocity. The x',y',z' reference frame moves with constant velocity V along the x axis of the x,y,z reference frame. At the time t, the displacement of the second reference frame with respect to the first is Vt.

reference frame with respect to the other is V, then the x-component of the velocity of the second reference frame with repect to the first is $v_x = V$, and the x-component of the velocity of the first with respect to the second is $v_{x'} = -V$.

We will call the x,y,z frame the "laboratory frame" and the x',y',z' frame the "spaceship frame." The motivation for this terminology is that in many instances one of the reference frames is at rest in the laboratory in which we are performing some experiment, and the second reference frame accompanies some high-speed body, moving through the laboratory at a speed that is perhaps close to the speed of light. More generally, we will call a reference frame that is attached to some body or accompanies some body the **rest frame** of that body. Thus, we speak of the rest frame of a train, or the rest frame of a proton, or the rest frame centered on the Earth.

If we measure the position coordinates and the velocity of a particle with respect to the laboratory frame and we simultaneously measure the position coordinates and the velocity of the same particle with respect to the spaceship frame, we will obtain different results--position coordinates and velocities depend on the choice of reference frame. However, the coordinates and velocities obtained in one reference frame are related to those obtained in the other reference frame by simple transformation formulas.

For the time coordinate t, the transformation formula is trivial. Since time is absolute in Newtonian physics, we can assume that the clocks in the spaceship frame are synchronized with those in the laboratory frame and, consequently, the values of the time coordinates measured in the two reference frames coincide,

$$\boxed{t = t'} \tag{1}$$

For the space coordinates x,y,z, the transformation formulas can be obtained by simple geometry, taking into account that in Newtonian physics lengths are absolute. We assume that at time $t = t' = 0$, the origins of the two reference frames coincide, and that consequently at a time $t = t' \neq 0$, the displacement of the origin O' of the spaceship frame relative to the origin O of the laboratory frame is Vt, in the x direction. Figure 1.6 shows this displacement, and it also shows the coordinates x,y and x',y' of a point P at this time. From this figure we see that the coordinate x is simply the sum of the coordinate x' and Vt,

$$\boxed{x = x' + Vt} \tag{2}$$

Equivalently, the coordinate x' is the difference between x and Vt,

$$x' = x - Vt \tag{3}$$

To obtain the result (2), we have added the distance x' defined in the spaceship frame to the distance Vt defined in the laboratory frame, and we have obtained a distance x defined in the laboratory frame. Such a mixing of distances defined in different reference frames is permissible in Newtonian physics because of the absolute character of length. The distance between two points at one instant of time is not affected by their position or by their motion, and when-

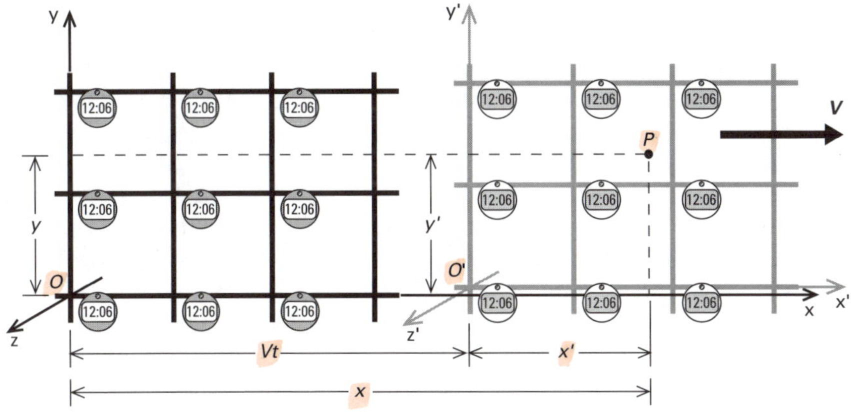

Fig. 1.6 The coordinates of the point P are x,y with respect to the laboratory frame and x',y' with respect to the spaceship frame (for the sake of simplicity, the z and z' coordinates are taken as zero in this diagram).

ever we measure the distance between two points at one instant of time in one reference frame, we obtain exactly the same result as if we had instead measured this distance in another reference frame. Thus, observers in the laboratory frame and in the spaceship frame agree on the measured value of the distance x' between O' and P, and this distance x' has the same value in both reference frames. Likewise, the distance Vt between O and O' has the same value in both reference frames. It is therefore correct to add x' and Vt to obtain the distance between O and P, and this sum is the value of the x coordinate in the laboratory frame, as stated in Eq. (2).

From Fig. 1.6 we also see that the coordinate y' coincides with y,

$$y' = y \tag{4}$$

This merely expresses the absolute character of distances measured along the y direction. Likewise, the coordinate z' (not shown in Fig. 1.6) coincides with z,

$$z' = z \tag{5}$$

Note that the three Eqs. (3) - (5) can be combined in a single vector equation. We will adopt the boldface letter $\mathbf{x}$ for the three-dimensional position vector with components (x, y, z), that is,

$$\mathbf{x} = x\mathbf{i} + y\mathbf{j} + z\mathbf{k} \tag{6}$$

where $\mathbf{i}$, $\mathbf{j}$, and $\mathbf{k}$ are the standard unit vectors along the x, y, and z axes. Then the vector form of Eqs. (3) - (5) is

$$\mathbf{x}' = \mathbf{x} - \mathbf{V}t \tag{7}$$

This can be verified by examining the x, y, and z components of the vector equation, keeping in mind that the velocity vector $\mathbf{V}$ is in the x direction.

Equation (1) and Eqs. (3) - (5) are the transformation equations that express the time coordinate and the space coordinates measured in one reference frame in terms of those measured in another reference frame. These equations are called the **Galilean transformation**. As we saw above, these equations hinge on the absolute character of time and of length, an assumption that lies at the foundations of Newtonian physics. The Galilean transformation is therefore valid as long as Newton's physics is valid, that is, as long as we are dealing with the speeds we encounter in our everyday experience, which are very small compared with the speed of light. In later chapters we will see how and why the Galilean transformation and Newtonian physics fail when the speeds are not very small compared with the speed of light.

The Galilean transformation expresses a geometrical relationship between the coordinates of one reference frame and those of another. Accordingly, these transformations can be expressed in a graphical form, in a spacetime diagram. To simplify the diagram, we will ignore the space coordinates y and z [which, according to Eqs. (4) and (5) have trivial transformation equations], and we will focus on the two coordinates x and t. The spacetime diagram then reduces to a graph of the space coordinate x and the time coordinate t.

Figure 1.7 is an example of such a simplified spacetime diagram, showing only the x axis, the t axis, and the rectangular grid that determines that x,t coordinates. The worldline shown in this diagram represents the motion of some particle relative to the laboratory frame (x, y, z frame) and the spacetime point P represents some event on this worldline. If we want to represent the motion of this particle relative to the spaceship frame (x', y', z' frame), we can proceed in one of two ways: we can draw a new spacetime diagram with a rectangular grid x',t' in which we plot the new worldline that represents of motion of the particle in these new coordinates, or we can keep the spacetime diagram of Fig. 1.7 with its already plotted worldline and draw a new, additional grid x',t' on top of it, such that the same worldline repre-

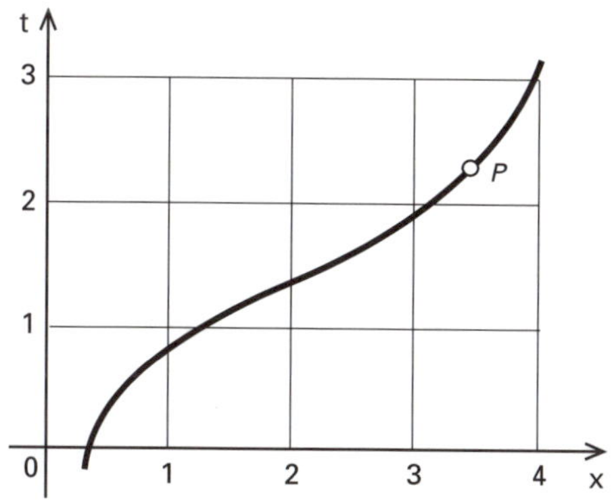

Fig. 1.7 A simplified spacetime diagram with only the x and t axes. The worldline shown represents the motion of some particle; P is an event on the worldline.

sents the motion relative to the new grid. The second procedure is preferable, since it emphasizes that what we are changing is the reference frame, not the motion of the particle or the spacetime points that represent events.

How must we draw the new grid x',t' on top of Fig. 1.7 so that the same worldline represents the motion of the particle relative to the spaceship? To discover this, we begin by plotting the worldline of the origin of the spaceship frame. This is simply the straight line $x = Vt$, slanted to the right, corresponding to the velocity V (see Fig. 1.8). All the points along this line have $x' = 0$, and only t' varies along this line. Thus, this slanted line must be the new t' axis. To discover the new x' axis, we note that this axis is the set of points that have $t' = 0$. But because of the absolute character of time, $t' = 0$ is the same as $t = 0$; consequently, the x' axis coincides with the x axis.

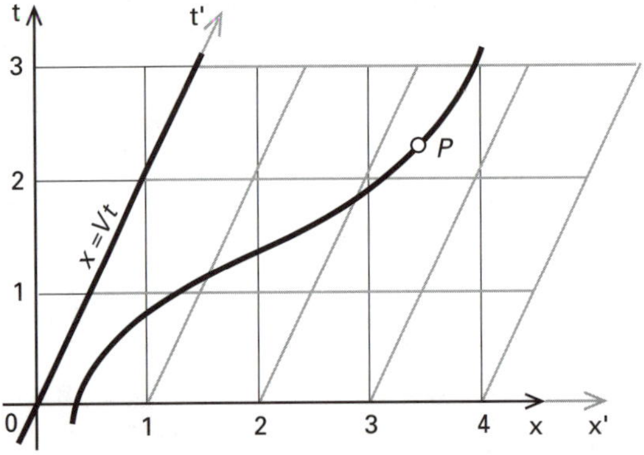

Fig. 1.8 Spacetime diagram with grid of x,t coordinates and grid of x',t' coordinates. The slanted line $x = Vt$ is the worldline of the origin of the spaceship frame.

Figure 1.8 shows the grid of the new x',t' coordinates on top of the old grid of x,t coordinates. This figure gives us a graphical representation of the Galilean transformation. The laboratory and the spaceship coordinates of any event P can be determined graphically by drawing intercepts toward the appropriate axes (see Fig. 1.9). Note that when using a slanted pair of axes, such as x',t', we must construct the intercepts with lines parallel to the axes. Thus, we must contstruct the intercept on the x' axis by drawing a slanted line parallel to the t' axis. From the intercepts displayed in Fig. 1.9, we immediately see that values of the laboratory and the spaceship coordinates are related as follows:

$$t' = t$$

$$x' = x - Vt$$

Thus, our graphical construction permits us to rederive the Galilean transformation equations.

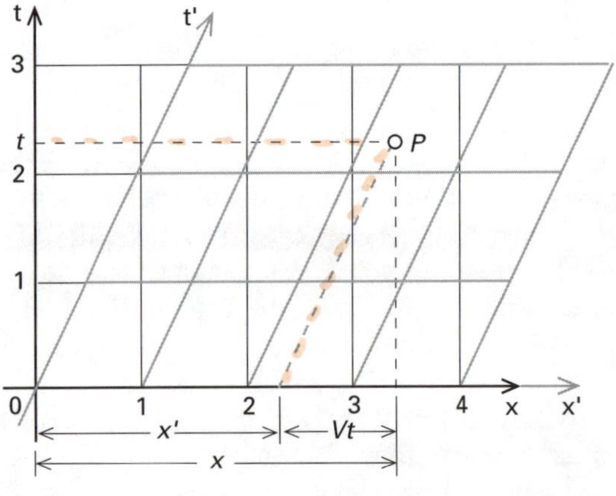

Fig. 1.9 The coordinates x,t and x',t' of an event P can be determined graphically by drawing the intercepts to the appropriate axes.

1.4 THE ADDITION OF VELOCITIES

It seems intuitively obvious that velocities measured in different reference frames should combine by simple addition. For instance, if two automobiles are traveling in opposite directions on the same road, one traveling east at 50 km/h and the other west at 60 km/h, then in the reference frame of the first automobile, the second should have a westward relative speed of 110 km/h. This simple addition rule for speeds can be shown to be a consequence of the Galilean transformation. To see this, we take the differentials of the coordinates in Eqs. (1) and (3):

$$dt' = dt \tag{8}$$

$$dx' = dx - V dt \tag{9}$$

Dividing the second equation by the first, we find

$$\frac{dx'}{dt'} = \frac{dx}{dt} - V \tag{10}$$

or

$$v'_x = v_x - V \tag{11}$$

This is a simple addition law for the x components of the velocity.

Since the y and z coordinates are not changed by the Galilean transformation, these components of the velocity are not changed:

$$v'_y = v_y \tag{12}$$

$$v'_z = v_z \tag{13}$$

The three equations for the transformation of the components of the velocity can be combined in a single vector equation:

$$\boxed{\mathbf{v}' = \mathbf{v} - \mathbf{V}} \tag{14}$$

Thus, the velocities $\mathbf{v}$ and $\mathbf{V}$ are combined by vector addition (or subtraction); accordingly, Eq. (14) is called the **Galilean addition law for velocity.**

In general, the transformation of the velocity from one frame to another changes not only the magnitude of the velocity but also the direction.

■ **EXAMPLE 2.** Raindrops are falling vertically with a speed of 5 m/s. A bicyclist races through this rain with a horizontal speed of 8 m/s relative to the ground. What are the magnitude and the direction of the velocity of the raindrops in the reference frame of the bicyclist?

SOLUTION. According to the addition law (14), the velocity $\mathbf{v}'$ of a raindrop in the bicyclist's reference frame is related to the velocity $\mathbf{v}$ in the ground frame by $\mathbf{v}' = \mathbf{v} - \mathbf{V}$. Figure 1.10 shows the vectors $\mathbf{v}$ and $\mathbf{V}$ and their difference. From the vector triangle, the magnitude of $\mathbf{v}'$ is

$$v' = \sqrt{v^2 + V^2} = \sqrt{5^2 + 8^2} \text{ m/s} = 9.4 \text{ m/s}$$

and the angle ϕ between the vector $\mathbf{v}'$ and the vertical is

$$\begin{aligned} \phi &= \tan^{-1} V/v \qquad (15) \\ &= \tan^{-1} 8/5 = 58^0 \end{aligned}$$

Thus, in the bicyclist's reference frame, the rain is tilted backward, striking her face. Figure 1.11a,b compares the velocity vectors of the raindrops in the two reference frames.

Fig. 1.10 The velocity vectors $\mathbf{v}$, $\mathbf{v}'$, and $\mathbf{V}$ form a vector triangle, according to $\mathbf{v}' = \mathbf{v} - \mathbf{V}$.

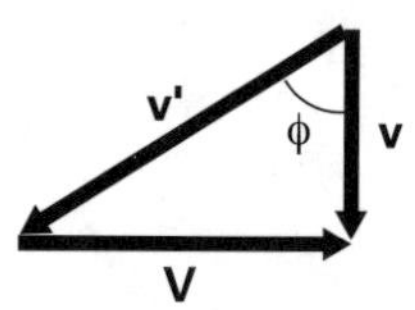

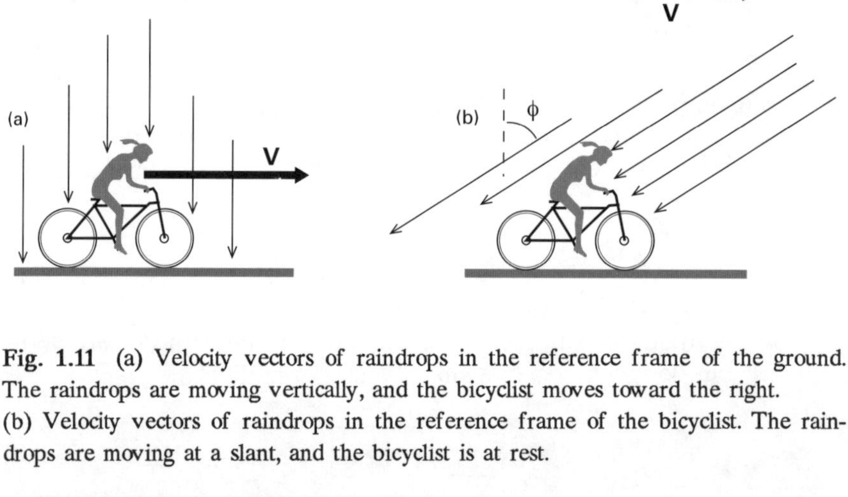

Fig. 1.11 (a) Velocity vectors of raindrops in the reference frame of the ground. The raindrops are moving vertically, and the bicyclist moves toward the right.
(b) Velocity vectors of raindrops in the reference frame of the bicyclist. The raindrops are moving at a slant, and the bicyclist is at rest.

From the transformation law for the velocity, we immediately obtain a transformation law for the acceleration. Since $\mathbf{V}$ is a constant, differentiation of Eq. (14) gives

$$d\mathbf{v}' = d\mathbf{v}$$

and dividing this by Eq. (8), we obtain

$$\frac{d\mathbf{v}'}{dt'} = \frac{d\mathbf{v}}{dt}$$

This simply says that the acceleration is unchanged, that is, the *acceleration is absolute*:

$$\boxed{\mathbf{a}' = \mathbf{a}} \tag{16}$$

Note that if the acceleration of a particle is zero in the first reference frame, then it also zero in the second. Thus, a particle that moves with uniform velocity with respect to the first reference frame also moves with uniform velocity relative to the second [although the magnitudes and directions of these uniform velocities will be different, as given by Eq. (14)]. This is exactly what is required to ensure the validity of Newton's First Law in both reference frames.

The absolute character of acceleration also ensures that Newton's Second Law is valid in both reference frames, that is, Newton's Second Law is unchanged by the transformation from one inertial reference frame to another. In the first inertial reference frame, Newton's Second Law is

$$m\mathbf{a} = \mathbf{F} \tag{17}$$

Substituting $\mathbf{a} = \mathbf{a}'$ and $\mathbf{F} = \mathbf{F}'$ in this equation, we find

$$m\mathbf{a}' = \mathbf{F}' \tag{18}$$

which is Newton's Second Law in the second reference frame. Note that here we have assumed that the force is unchanged by the transformation of reference frame, that is, force is absolute. To justify this assumption, we need to recall how the force appearing on the right side of Newton's Second Law is measured. In principle, before the force $\mathbf{F}$ is applied to the mass m, the force can be measured by a preliminary procedure that involves applying this same force to a standard test particle or test body. For instance, we can apply the force to the standard kilogram, and we can then calculate the force from the resulting acceleration of the standard kilogram. This makes it clear that the absolute character of force is a direct consequence of the absolute character of acceleration.

The validity of Newton's Second Law in all inertial reference frames is in accord with Galileo's ideas about the relativity of uniform motion--since the law of motion is the same in all reference frames, the mechanical phenomena we observe in one reference frame and another will be the same. More generally, we find that all the laws of mechanics are the same in all inertial reference frames. We say that the laws of mechanics are **invariant** under transformations from one inertial reference frame to another, and we call this invariance the **Principle of Relativity for Mechanics**, or the **Galilean Principle of Relativity.**

Even though all the laws of mechanics are the same in all inertial reference frames, the values of some of the mechanical quantities suffer changes when we transform from one reference frame to another. We have already seen in Eq. (14) how the velocity transforms. As a consequence of this transformation, the momentum of a particle changes:

$$\mathbf{p}' = m\mathbf{v}' = m(\mathbf{v} - \mathbf{V}) \tag{19}$$

that is,

$$\mathbf{p}' = \mathbf{p} - m\mathbf{V} \tag{20}$$

Likewise, the kinetic energy changes:

$$K' = \frac{1}{2}m\mathbf{v}'^2 = \frac{1}{2}m(\mathbf{v} - \mathbf{V})\cdot(\mathbf{v} - \mathbf{V}) = \frac{1}{2}m(v^2 + V^2 - 2\mathbf{v} \cdot \mathbf{V}) \tag{21}$$

That is,

$$K' = K + \frac{1}{2}mV^2 - \mathbf{p} \cdot \mathbf{V} \tag{22}$$

Note that although these transformation laws change the values of the momentum and of the energy, these changes do not conflict with the conservation of momentum and of energy in elastic collisions. The term $m\mathbf{V}$ on the right side of Eq. (20) is merely a constant, and if the sum of the momenta $\mathbf{p}$ for a system of particles is constant, then the sum of the momenta $\mathbf{p}'$ will also be constant. Likewise, the term $\frac{1}{2}mV^2$ on the right side of Eq. (22) is a constant, and, by momentum conservation, the sum of the terms $\mathbf{p} \cdot \mathbf{V}$ is a constant for a system of particles without external forces. Hence, if the sum of the kinetic energies K is constant for a system of particles without external forces, then the sum of the kinetic energies K' will also be constant.

1.5 THE WAVE THEORY OF LIGHT; THE ETHER

In the preceding sections we saw that Newtonian mechanics is in accord with the Galilean Principle of Relativity--the laws of mechanics are the same in all inertial reference frames, and there are no mechanical experiments that reveal any intrinsic distinction between one inertial reference frame and another. Thus, the translational motion of the spaceship frame with respect to the laboratory frame cannot be detected by experiments within the spaceship frame: it can be detected only by observing some landmark located in the laboratory frame. As far as mechanics is concerned, all motion is relative.

Before the days of Einstein, physicists understood this relativity of mechanics, but they believed that relativity did not apply to the propagation of light and that it did not apply to some electric and magnetic experiments. Early in the 19th century, the experimental and theoretical investigations of Thomas Young and Augustin Fresnel had established that light is a wave. At first, it was thought that light was a mechanical wave consisting of undulations in a tenuous elastic medium that was supposed to pervade all of space. The propagation of light in this medium was thought to be analogous to the propagation of sound in air. The hypothetical medium that served as the carrier of light waves was called the **ether** (in earlier centuries, the word *ether* was used to designate a ghostly substance that was believed to fill the upper regions of the universe beyond the orbit of the Moon). Besides filling empty space, the ether was also supposed to permeate all materials and to coexist with them, so as to permit the propagation of light in air, water, glass, etc. However, the ether was supposed to have no direct effect on the materials immersed in it and no effect on the motion of planets, particles or anything else--its only purpose was to serve as medium for the undulations of light waves. As a witticism expressed it, "The ether is merely the subject of the verb to undulate."

In the reference frame in which the ether is at rest, or the ether frame, light waves have a speed $c = 3 \times 10^8$ m/s,* the same in all directions. But according to the Galilean addition rule for velocities, in any other reference frame moving at some velocity $\mathbf{V}$ relative to the ether, the velocity of light is then $\mathbf{c} - \mathbf{V}$. The magnitude of this velocity is larger or smaller than 3×10^8 m/s, and the magnitude depends on direction. This establishes an absolute distinction between reference frames: there is one and only one preferred reference frame in which the speed of light is 3×10^8 m/s, and in all other reference frames the speed of light is either larger or smaller. Therefore, measurements of the speed of light would permit us to detect the motion of our reference frame through the ether. This would imply that velocities are absolute--the ether provides us with a universal background for velocity measurements, and the preferred rest frame of the ether establishes an absolute standard of rest.

The theoretical physicists of the 19th century spent much effort on developing mechanical models of the ether that were intended to

---

* Here and elsewhere, we round off the value of the speed of light for the sake of convenience. The exact value of the speed of light is $c = 2.99792458 \times 10^8$ m/s.

account for the transverse polarization of light, with its perpendicular electric and magnetic fields. But these models were abandoned after James Clerk Maxwell discovered the electromagnetic character of light. When Maxwell formulated his equations for electricity and magnetism, he found that these equations predicted a mutual induction of electric and magnetic fields, which could maintain oscillating electric and magnetic fields in empty space. These self-supporting electromagnetic fields had the character of a wave propagating with a definite speed of $1/\sqrt{\mu_0\epsilon_0} = 3 \times 10^8$ m/s. The agreement between this theoretical speed and the experimentally measured speed of light led Maxwell to conclude that light is such an electromagnetic wave. The recognition that light consists of oscillating electric and magnetic fields led to the rejection of mechanical models of the ether, but it did not eliminate the requirement for a preferred reference frame in which the speed of light has the special value 3×10^8 m/s. The preferred reference frame continued to be called the ether frame, even though the idea of a mechanical ether gradually faded into history (not all of Maxwell's contemporaries were willing to abandon the mechanical ether, and the idea lingered well into the 20th century). In this preferred reference frame, Maxwell's equations have their simple, familiar form. But in all other reference frames, Maxwell's equations are more complicated, since, according to the ether theory, they have to be constructed from the simple equations by applying the Galilean transformation.

1.6 THE MICHELSON-MORLEY EXPERIMENT

We can regard the motion of the ether relative to our reference frame as an ether wind, and we can then think of the propagation of a light wave as analogous to the propagation of a sound wave in air on a windy day. To the 19th-century physicists, it seemed plausible that the Sun would be at rest in the ether, so the ether wind blowing through the reference frame of the Earth would have a velocity opposite to that of the Earth around the Sun, about 30 km/s. The speed of light in the reference frame of the Earth would then depend on direction: the light wave would have a maximum speed when propagating downwind, a reduced speed when propagating across the wind, and a minimum speed when propagating upwind. According to the Galilean addition law, if the velocity of the Earth relative to the ether is $\mathbf{V}$, then the velocity of a light wave relative to the Earth is $\mathbf{c} - \mathbf{V}$. The magnitude of this velocity is $c + V$ for a light wave in the downwind direction, $c - V$ for the upwind direction, and $\sqrt{c^2 - V^2}$ for the direction across the wind (see Fig. 1.12). With $V \simeq 30$ km/s, the fractional change of the speed of the light wave in relation to its standard speed c is about $\pm 10^{-4}$ for the downwind and upwind directions, respectively, and about 0.5×10^{-8} for the direction across the wind. However, the 19th-century techniques for measuring the speed of light required that the light wave complete a round trip, and for such a round trip the upwind and downwind speed changes tend to average out, leaving only a small observable residue of about 10^{-8}, which is about the same as for a trip across the wind.

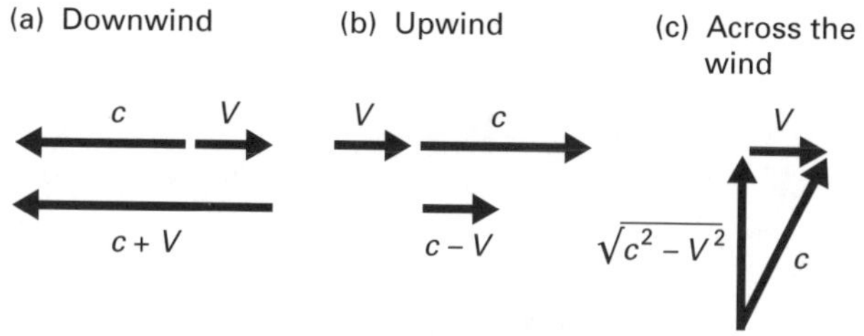

Fig. 1.12 The velocity of a light wave relative to the Earth is the vector difference **c** - **V**. This vector difference has a magnitude $c + V$ for a light wave in the downwind direction (a), $c - V$ for the upwind direction (b), and $\sqrt{c^2 - V^2}$ for the direction across the wind (c).

The earliest attempts at detecting the ether wind were made by M. Hoek in 1868 and by G. B. Airy in 1871, who investigated the propagation of light in water-filled tubes and astronomical telescopes. They found no effect, but this result was inconclusive because according to a version of the ether theory formulated by A. Fresnel, no effect was expected in the Hoek and Airy experiments. Fresnel had proposed that transparent materials, such as water, partially drag the ether along, and he had proposed the Hoek experiment as a means of verifying this ether drag.

The first conclusive experiment capable of detecting the small difference between the parallel and the perpendicular speeds of light was devised by A. A. Michelson in 1881. A few years later, he repeated it in an improved form in cooperation with E. W. Morley, and it became known as the **Michelson-Morley experiment** [3]. The experiment uses an interferometer with two perpendicular arms of equal lengths (MM_1 and MM_2), with a half-silvered mirror M at the intersection of the arms and fully silvered mirrors M_1 and M_2 at the ends of the arms (see Fig. 1.13). A light beam entering the interferometer is split by the half-silvered mirror into two separate beams that travel outward and back along the two arms and then emerge at A, where they are allowed to interfere. A difference of travel time along the two arms results in a phase difference between the two beams. Depending on this phase difference, the interference will be constructive or destructive, and the experimenter looking toward the beams detects this as brightness or darkness in the center of his field of view.

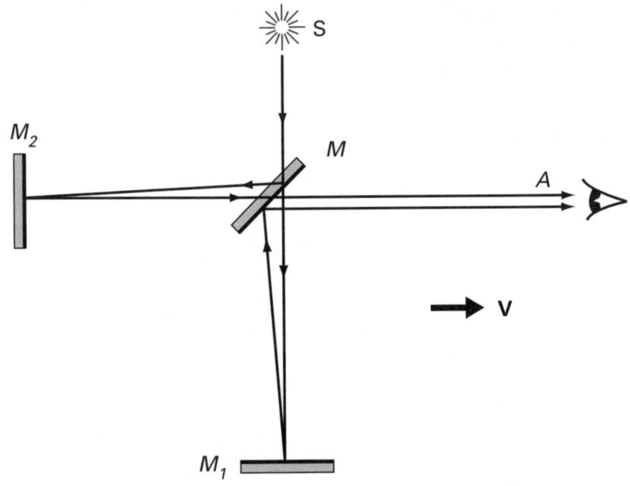

Fig. 1.13 Schematic diagram of the Michelson-Morley interferometer.

We can express the phase difference between the beams in terms of the magnitude of the ether wind as follows: Suppose that the arm MM_1 is perpendicular to the ether wind, and MM_2 is parallel to it. The light beam traveling along the arm MM_1 has a direction across the ether wind and hence a speed $\sqrt{c^2 - V^2}$. If the length of the arm is L, then the time for the round trip of this light beam is

$$t_1 = \frac{2L}{\sqrt{c^2 - V^2}} = \frac{2L/c}{\sqrt{1 - V^2/c^2}} \tag{23}$$

The light beam traveling along the arm MM_2 has a downwind direction (toward the left in Fig. 1.13) during the outward part of the trip and an upwind direction (toward the right) during the return part of the trip; hence it has a speed $c + V$ during the outward part of the trip and a speed $c - V$ during the return part. The time for the round trip of this light beam is therefore

$$t_2 = \frac{L}{c + V} + \frac{L}{c - V} = \frac{L}{c}\left(\frac{1}{1 + V/c} + \frac{1}{1 - V/c}\right) = \frac{2L/c}{1 - V^2/c^2} \tag{24}$$

If the wavelength of the light is λ, the phase difference between the emerging light beams is

$$\begin{aligned} \Delta\phi &= \frac{2\pi c}{\lambda}(t_2 - t_1) \\ &= \frac{2\pi}{\lambda} L \left(\frac{2}{1 - V^2/c^2} - \frac{2}{\sqrt{1 - V^2/c^2}}\right) \end{aligned} \tag{25}$$

Since $V << c$, we can use the approximations

$$\frac{1}{1 - V^2/c^2} \simeq 1 + \frac{V^2}{c^2} \quad \text{and} \quad \frac{1}{\sqrt{1 - V^2/c^2}} \simeq 1 + \frac{V^2}{2c^2}$$

so the phase difference becomes

$$\Delta\phi = 2\pi \frac{L}{\lambda} \left(\frac{2V^2}{c^2} - \frac{2V^2}{2c^2} \right)$$

$$= 2\pi \frac{L}{\lambda} \frac{V^2}{c^2} \tag{26}$$

Thus, the ether wind V causes a phase difference between the beams, and they will interfere constructively or destructively, depending on the magnitude of this phase difference.

This phase difference can be conveniently detected by rotating the entire interferometer by 90°, thereby reversing the roles of the parallel and perpendicular arms. During this rotation, the phase difference gradually decreases from the value (26) to its negative, and the interference of the light beams alternates from constructive, to destructive, to constructive, etc. These changes in the interference serve as a sensitive test for the existence of an ether wind.

The interferometer of Michelson and Morley (see Fig. 1.14) was built on a massive stone slab floating in a pool of mercury, so it could by easily and smoothly rotated by 90°. Furthermore, the interferometer had multiple mirrors at the ends of its arms, so that the light beams would travel back and forth along the arms several times. This increased the effective arm length L to 11 m and gave the instrument a sensitivity adequate to detect an ether wind as small as 4.7 km/s. But Michelson and Morley failed to detect any ether wind whatsoever with their interferometer. Since the expected wind is about 30 km/s, their null result contradicted the ether theory of the propagation of light. As we will see in the next chapter, this null result is in perfect agreement with Einstein's theory of relativity.

■ **EXAMPLE 3.** With an effective length L = 11 m and with λ = 590 nm, what phase shift will result from a rotation of the interferometer by 90° in an ether wind of 30 km/s? Michelson and Morley claimed they could observe a phase shift as small as 0.06 radian. If so, what ether wind could they detect?

SOLUTION. For a rotation by 90°, the phase shift is twice that given in Eq. (26):

$$\Delta\phi = 4\pi \frac{L}{\lambda} \frac{V^2}{c^2} = 4\pi \frac{11\text{ m}}{590 \times 10^{-9}\text{ m}} \left(\frac{30\text{ km/s}}{3 \times 10^5\text{ km/s}} \right)^2 = 2.3\text{ radians}$$

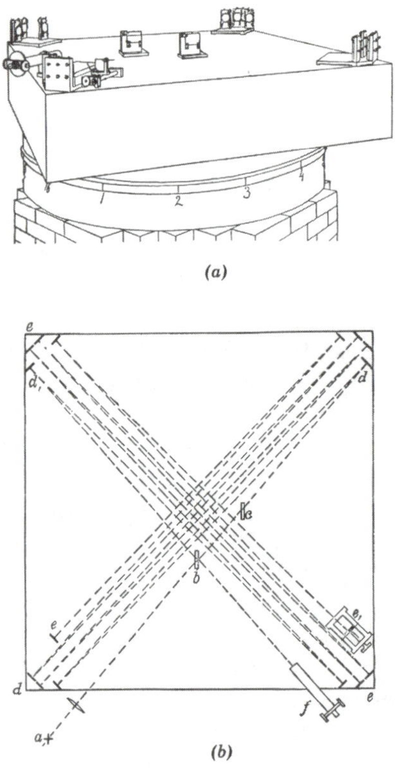

Fig. 1.14 The interferometer of Michelson and Morley was mounted on a large stone slab (a), and it had multiple mirrors, which reflected the light back and forth several times (b). [From A. A. Michelson and E. W. Morley, Phil. Mag. **190**, 449 (1887)]

The minimum observable phase shift of 0.06 radian is 39 times smaller than 2.3 radians, hence it corresponds to a value of V/c that is $\sqrt{39} = 6.25$ times smaller. The minimum detectable ether wind is therefore (30 km/s)/6.25 = 4.8 km/s.

Because of its crucial importance in discriminating between Newtonian physics and relativistic physics, the Michelson-Morley experiment was repeated many times throughout the 20th century. Table 1.1 summarizes the results of all these repetitions. No ether wind was detected in any of these experiments. The last column in the table gives the upper limit V_{eff} for the speed of the ether wind consistent with the experimental errors. The most precise interferometer experiment was that by Joos in 1930, which set a limit of 1.5 km/s on the ether wind.

TABLE 1.1 MICHELSON-MORLEY EXPERIMENTS*

Experimenters	Year	V_{eff}
Michelson	1881	21 km/s
Michelson and Morley	1887	4.7
Morley and Miller	1902-04	3.3
Miller	1921	7.7
Miller	1923-24	7.5
Miller (sunlight)	1924	3.3
Tomashek (starlight)	1924	7.7
Miller	1925-26	8.3
Kennedy	1926	5.0
Illingworth	1927	2.2
Piccard and Stahel	1927	6.8
Michelson et al.	1929	3.1
Joos	1930	1.5
Essen (frequencies of microwave cavities) [5]	1955	2.5
Jaseja et al. (frequencies of lasers) [6]	1964	0.95
Brillet and Hall (frequencies of etalons) [7]	1979	0.015

* The entries before 1930 are based on a table by Shankland et al. [4]

The more recent Michelson-Morley experiments listed in Table 1.1 did not compare the the phases of light beams traveling at right angles but, rather, the frequencies of standing waves in resonant cavities oriented at right angles. Such a standing wave can be regarded as a traveling wave bouncing back and forth between the ends of the cavity, and the frequency of the standing wave is directly proportional to the speed of the traveling wave. If the resonant cavities are of the same length, then any observed difference in their frequencies must be attributed to a difference in the speeds of the light waves. The frequency difference, if any, can be accurately detected by superposing waves radiated from the two cavities; the frequency difference will then cause beats in the combined wave. The last three experiments listed in Table 1.1 relied on such frequency comparisons, but they differed in the details of their experimental techniques. Essen [5] used resonant microwave cavities; Jaseja et al. [6] used lasers as optical resonant cavities; and Brillet and Hall [7] used Fabry-Perot etalons (paired parallel mirrors, with light bouncing back and forth) as optical resonant cavities. The last experiment set a remarkably small limit of 0.015 km/s, or 15 m/s, on the ether wind.

The null result of the Michelson-Morely experiment was viewed with alarm by 19th-century physicists. They proposed several explanations of this null result. One of these explanations was a modified ether-drag theory, according to which a large, heavy body, such as the Earth, drags along not only the ether in its interior but also the ether in its immediate vicinity. This induced drag eliminates the ether wind in laboratories on the surface of the Earth. This was the explanation favored by Michelson himself, who regarded the outcome of his experiment as confirmation of ether drag. However, the notion of ether drag is in conflict with stellar **aberration**, that is, the apparent seasonal shift of the observed positions of the stars in the sky. This shift, which can be as large as 40 arcseconds, is caused by the orbital velocity of the Earth. It is analogous to the tilt of the trajectories of raindrops when observed by a runner or a bicyclist racing through a rain shower, as discussed in Example 2. Stellar aberration would be absent if the ether in the vicinity of the Earth were dragged along. Furthermore, if the translational motion of the Earth drags the ether along, then the rotational motion of the Earth should do likewise. But this is not the case, since the rotational motion of the Earth is detectable with optical experiments, as was first demonstrated by Michelson and Gale [8] in 1925 by an experiment with two light beams traveling in opposite directions around a closed path defined by mirrors.

Another explanation was proposed by G. F. Fitzgerald and by H. A. Lorentz, who postulated that whenever a body moves through the ether at speed V, it suffers a length contraction along the direction of motion by a factor of $\sqrt{1 - V^2/c^2}$. This means that the arm of the interferometer parallel to the direction of motion suffers this length contraction, but the perpendicular arm does not. Accordingly, we must multiply the length L in Eq. (24) by $\sqrt{1 - V^2/c^2}$, which has the consequence that the travel times t_1 and t_2 become equal, and the phase difference (25) becomes zero. Although the contraction hypothesis explained the null result obtained by Michelson and Morley, a modified experiment by Kennedy and Thorndike [9], employing an interferometer with arms of unequal length, gave evidence against the contraction hypothesis. If the arms of the interferometer are unequal, say, L_1 and L_2, then the phase difference is nonzero, but the contraction makes the phase difference independent of the orientation of the arms, and hence rotation of the instrument has no observable effect. However, the phase difference contains a term proportional to V^2 (see Problem 19),

$$\Delta\phi = [\text{const.}] + \frac{2\pi}{\lambda}(L_2 - L_1)V^2 \tag{27}$$

and this results in diurnal and seasonal variations. Kennedy and Thorndike performed their experiment in 1932, and by then it was known that the Solar System orbits around the center of the Galaxy at a speed of about 250 km/s; it was therefore expected that the ether wind would have this magnitude relative to the Sun. Between midnight and noon, the rotational motion of the Earth changes the forward velocity of a point on the equator by about $\Delta V \simeq 1$ km/s, and between one time of the year and a time six months later, the orbital motion of the Earth changes the velocity of the Earth by about $\Delta V \simeq$ 60 km/s. Thus, the expected diurnal variation in V^2 is about $\Delta V^2 \simeq 2V\Delta V \simeq 2 \times 250$ km/s $\times$ 1 km/s = 250 (km/s)^2, and the expected seasonal variation is about $\Delta V^2 \simeq 2V\Delta V \simeq 2 \times 250$ km/s $\times$ 60 km/s $\simeq$ 15,000 (km/s)^2. These large variations in V^2 would produce easily detectable variations in the phase shift. But Kennedy and Thorndike detected no such variations, and from their experimental errors they concluded that the observed ether wind associated with the motion of the Solar System is not 250 km/s, but at most 15 km/s. This null result demonstrates that the Fitzgerald-Lorentz contraction is a failure--it does not achieve its aim of removing the discrepancy between observation and the ether theory of the propagation of light.

Recent versions of the Kennedy-Thorndike experiment have set better limits on the Solar System ether wind and confirmed the failure of the Fitzgerald-Lorentz contraction (see Table 1.2). Hils and Hall [10] used a Fabry-Perot etalon as an optical resonant cavity and compared the resonant frequency of this cavity with the frequency of an atomic standard. The period of the standing wave in such a cavity is given by the travel time of light bouncing back and forth between its ends, that is, the period is given by Eq. (23) [or, equivalently, by Eq. (24) with an extra length-contraction factor]. Thus, the period depends on V^2, and it is subject to diurnal and seasonal variations. The experimenters checked for diurnal variations, and from the absence of any such variations they set a limit of 50 m/s on the ether wind. Müller and Soffel [11] exploited the available data from lunar laser ranging, that is, measurements of the travel time of laser pulses sent to the Moon and reflected back to Earth by corner reflectors installed on the Moon by the Apollo 11 astronauts. If the effective speed of light for a round trip is $\sqrt{c^2 - V^2}$, then the speed has seasonal variations, and the travel time will fluctuate seasonally. From the absence of such seasonal fluctuations, the investigators were able to set a limit of 110 m/s on the ether wind.

TABLE 1.2 KENNEDY-THORNDIKE EXPERIMENTS

Experimenters	Year	V_{eff}
Kennedy and Thorndike [9]	1932	15 km/s
Hils and Hall [10]	1990	50 m/s
Muller and Soffel [11]	1995	110 m/s

Yet another explanation of the null result was attempted with emission theories, or ballistic theories, according to which light has the speed c not with respect to an ether frame but with respect to its source. This makes the emission of light by a source analogous to the firing of bullets from a gun. Accordingly, the speed of light relative to the observer depends on the motion of the source relative to the observer: if the source is at rest relative to the observer, the speed of light is c; if the source is moving toward the observer with speed V, it is $c + V$; and if the source is moving away, it is $c - V$. Since the original Michelson-Morley experiment used a light source at rest relative to the interferometer, no effect is expected. However, some of the later versions of the experiment were performed with sunlight and with starlight, and then an effect is expected, but none was observed. Furthermore, astronomers quickly recognized that a dependence of the speed of light on the speed of the source would lead to the observation of weird effects in binary systems of stars. In such systems, the two stars that are the sources of the light have different velocities in their orbits around each other. This would lead to different speeds for their light, and to different travel times for this light to reach the Earth. Consequently an astronomer on the Earth would see these stars as they were at different times in the past, which would result in an irregular behavior of their apparent orbital motion. But no such irregularities are observed, and this given us strong evidence against emission theories (we will discuss the observational evidence in Section 2.5).

The failure of all explanations of the null result of the Michelson-Morley experiment indicates that we need to question the foundations of Newtonian physics. As we will see in the next chapter, we have to abandon the absolute character of time and of length, and we have to abandon the Galilean transformation equations and the addition law for velocities.

*1.7 OTHER RECENT EXPERIMENTS**

The Michelson-Morley experiment is called a "second-order" experiment, because the observed effect depends on the square of the speed of the ether wind. We can see this from Eq. (26), where the phase difference is proportional to V^2. For many years it was thought that in any experiment in a laboratory on the Earth, the detectable effect of the ether wind would have to be of second order, since a closed path is required if the light is to interfere with itself. This means that the detectable effect always involves an average of the upwind and the downwind speeds, which leaves only a second-order residue, as in the Michelson-Morley experiment. But later it was rec-

* This section is optional.

ognized that experiments could be performed involving the frequency of light instead of interference, and the observable effect in such experiments can be of first order. (Note that the frequency experiments included in Table 1.1 are *not* of first order; they involve the average upwind and downwind speeds, and they are of second order, like the original Michelson-Morley experiment.)

The first-order frequency experiments rely on measurements of the Doppler shift of light from a fast-moving source. Since the propagation of light through the ether is analogous to the propagation of sound through air, the formula for the Doppler shift of light in Newtonian physics can be adapted from the familiar formula for sound. If the frequency emitted by the source is ν_S and the frequency observed at the receiver is ν_R, then the Doppler-shift formula for a receiver moving along a radial line directly toward or away from the source is

$$\nu_R = \nu_S \frac{1 - V_R/c}{1 - V_S/c} \tag{28}$$

Here the source velocity V_S and the receiver velocity V_R are reckoned relative to the ether; they are positive if in the direction of the outward radial line from the source, and negative otherwise. For a source and receiver moving along parallel lines at an angle θ with the direction of propagation of the light (see Fig. 1.15), the Doppler-shift formula has the generalization

$$\nu_R = \nu_S \frac{1 - (V_R/c)\cos\theta}{1 - (V_S/c)\cos\theta} \tag{29}$$

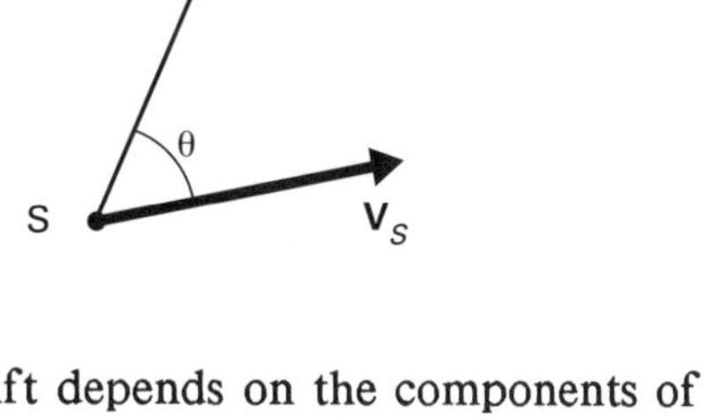

Fig. 1.15 A source of light (S) and a receiver (R) moving along parallel lines at an angle θ with respect to the direction of propagation of the light.

This simply says that the Doppler shift depends on the components of the velocities along the direction of propagation, not on the total velocities. Since $V_S/c << 1$, we can approximate Eq. (29) as

$$\nu_R \simeq \nu_S \left[1 - \frac{V_R - V_S}{c} \cos\theta\right] \tag{30}$$

The difference $V_R - V_S$ is the same in the laboratory frame and in the ether frame; this difference is simply the velocity of the receiver relative to the source. Hence it would seem that Eq. (30) cannot give us any information about the ether wind. However, the ether wind enters into Eq. (30) indirectly, because the angle of propagation θ in the ether frame differs from the angle of propagation in the laboratory frame by an amount that depends on the ether wind (this is analogous to the tilt of the trajectories of raindrops discussed in Example 2).

The maser experiment performed by Cedarholm et al. [12] used a molecular beam with an eastward velocity V_S as source and a second parallel molecular beam with a westward velocity V_R as receiver. Microwaves emitted perpendicularly from the source beam travel toward the "receiving" beam, where their frequency is compared with the frequency of the microwaves generated by the "receiving" beam. If the maser were at rest relative to the ether, then the direction of propagation would be perpendicular in both the laboratory frame and the ether frame (Fig. 1.16a). However, if the laboratory and the maser move with velocity V toward the east relative to the ether frame, then the direction of propagation is perpendicular in the laboratory frame but not in the ether frame. The direction of propagation in the ether frame is tilted slightly forward (see Fig. 1.16b), so the light travels at an angle of about $\phi = \tan^{-1} V/c \simeq V/c$ away from the perpendicular [see Eq. (15)]. Hence $\cos\theta = \cos(90^\circ - V/c) = \sin V/c \simeq V/c$, and Eq. (30) becomes

$$\nu_R \simeq \nu_S [1 - (V_R - V_S)V/c^2] \tag{31}$$

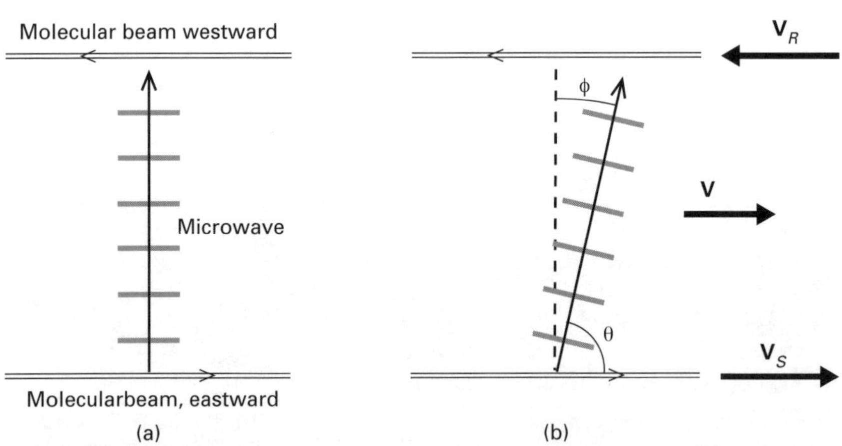

Fig. 1.16 Emission of microwaves by the molecular beam of a maser. (a) If the maser is at rest relative to the ether, then the direction of emission of the microwaves is at right angles to the beam. (b) If the maser has a velocity **V** toward the right, then the direction of propagation as seen in the ether frame is tilted forward at an angle $\phi \simeq V/c$.

From this we see that the frequency difference between the emitting molecules and the "receiving" molecules is proportional to $(V_R - V_S)V/c^2$, which is of first order in V. This dependence on the first order of V gives the Doppler-shift experiment a sensitivity higher than that of the Michelson-Morley experiment. Obviously, to enhance the sensitivity of the experiment, it is desirable to make $V_R - V_S$ as large as possible, which is done by using two high-speed molecular beams (with opposite directions of motion). In the maser experiment no Doppler shift was observed, and this null result gave a limit of 30 m/s for the ether wind. A somewhat similar experiment by Riis et al. [17] with an atomic beam of higher speed gave a limit of about 1 m/s.

Even better limits have been attained by performing Doppler-shift experiments with gamma rays emitted by a sample of radioactive nuclei acting as source, and absorbed resonantly by another sample of nuclei acting as receiver. The experiment exploits the Mössbauer effect; it was first performed by Hay et al. [13] and was repeated several times by other experimenters [14-16]. Figure 1.17 shows the experimental arrangement used by Champeney et al. [14]. The gamma rays were emitted by a sample of radioactive ^{57}Fe nuclei mounted on

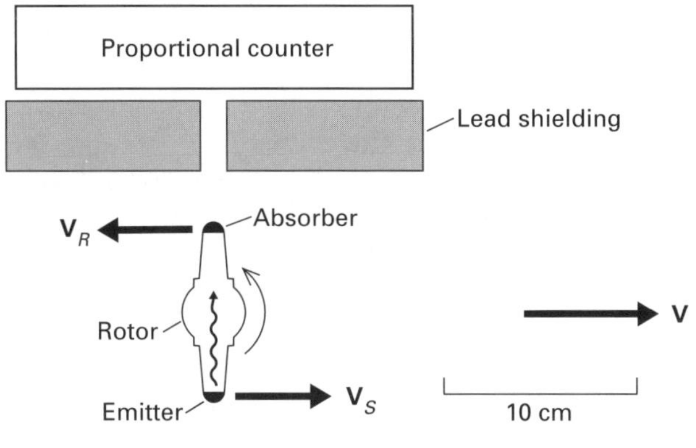

Fig. 1.17 An emitter and an absorber of gamma rays are mounted on the arms of a rapidly spinning rotor. At the instant shown, the direction of emission (aimed at the detector at the top) is at right angles to the direction of the velocity **V**.

the rim of a rapidly spinning rotor, and they were absorbed by a similar sample mounted at the diametrically opposite point of the rim. When emission and absorption proceed via the Mössbauer effect, the iron nuclei do not recoil during emission or absorption because they are "locked" to the crystal lattice (more details of the Mössbauer effect will be discussed in Section 5.6). Since the gamma rays do not lose any energy in recoils, the gamma rays emitted by the source are capable of triggering resonant absorption in the receiver. However, if the emitter and the absorber are in motion and the gamma rays suffer a Doppler shift as given by Eq. (31), then the mismatch of the source and receiver frequencies ν_S and ν_R would inhibit the resonant absorption of the gamma rays. This inhibition is a sensitive indicator of the presence of the Doppler shift. No inhibition of the absorption was detected, to within the experimental errors, and this placed a limit of 1.6 m/s on the ether wind. A later, improved version of the experiment by Isaak [16], lowered the limit to 5 cm/s, which is the tightest limit that has ever been set on the ether wind (see Table 1.3).

TABLE 1.3 DOPPLER-SHIFT EXPERIMENTS

Experimenters	Year	V_{eff}
Cedarholm et al. (molecular beam, maser) [12]	1958	30 m/s
Hay et al. (Mossbauer effect) [13]	1963	10
Champeney et al. (Mossbauer effect) [14]	1963	1.6 ± 2.8
Turner and Hill (Mossbauer effect) [15]	1963	2.2 ± 8.4
Isaak (Mossbauer effect) [16]	1970	0.05
Riis et al. (atomic beam, laser) [17]	1988	0.9

The development of highly accurate atomic clocks has made feasible yet another type of first-order experiment: the direct measurement of the one-way speed of light over a given distance. These measurements have set limits on the ether wind comparable to those obtained by frequency experiments (see Table 1.4).

The first application of this method was a by-product of Very Long Baseline Interferometry (VLBI) in radioastronomy. In this technique, astronomers use two or more radiotelescopes separated by thousands of kilometers to observe simultaneously a radio source in the sky. They make a digital recording of the radio waves incident on each radiotelescope and tag this recording with precise time signals from a synchronized clock located at each radiotelescope. They can later play back these synchronized recordings, mix them, and generate interference patterns to determine the location of the radio source in the sky. The time delay and the consequent interference between the radio waves arriving at two separate radiotelescopes depend on the extra path length that the radio wave has to travel (see Fig. 1.18), but they also depend on the speed of the wave. If this speed is altered by the component of an ether wind along the direction of travel of the radio wave, then successive determinations of the position of a radio source will display inconsistencies, since the relevant component of the ether wind will be different at different times of the year (for instance, for a radio source in the plane of the Earth's orbit, the wave speed is $c + V$ when the Earth is moving toward the source, it is $\sqrt{c^2 - V^2}$ three months later, and it is $c - V$ six months later). According to Cole [18], the available data indicate that the determinations of

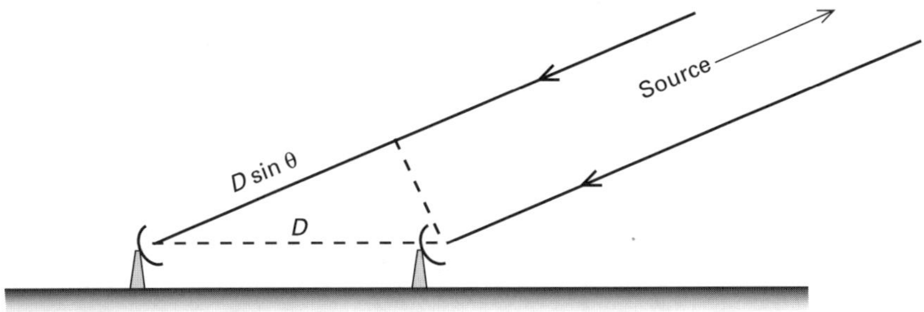

Fig. 1.18 Radio signal from a distant source in the sky. The signal has to travel an extra distance $D \sin\theta$ to reach the more distant radiotelescope.

the positions of radio sources at different times of the year are self-consistent to within 0.05 arcsec, and this implies that a limit of about 0.07 km/s on the ether wind.

A similar limit was obtained by Krisher et al. [19], who used a pair of synchronized hydrogen-maser clocks to check for a directional variation of the one-way speed of light traveling on an optic fiber between two stations separated by 21 km. Although the two stations were fixed on the Earth, they rotate with the Earth, and this changes the orientation of the path of the light relative to the ether, if any. The experimenters found no effect during observations extending over several days.

A much better limit was recently obtained by Wolf and Petit [20] from an analysis of data on travel times of radio signals transmitted by the satellites of the Global Positioning System (GPS). These satellites carry atomic cesium or rubidium clocks, synchronized to within about a nanosecond. The typical separation of a satellite from one of several ground stations equipped with their own synchronized clocks is about 20,000 km. Hence the one-way speed of a radio signal from the satellite to a ground station can be measured to within a fractional error of about $\delta c/c \simeq \delta t/t \simeq 10^{-9}\ \text{s}/(2 \times 10^{7}\ \text{m}/c) \simeq 3 \times 10^{-8}$. But this error can be reduced by averaging the data from different satellite-ground station pairs and from repeated passes of the satellites, for a chosen orientation of the path of the radio wave in space (relative to the fixed stars). The error can be reduced further by selectively extracting data that has the angular dependence expected for an ether wind. The net result is a reduction of the fractional error to about 2×10^{-9}, which corresponds to about 0.6 m/s. To within this observational error, no dependence of the speed of the radio waves on orientation was found.

TABLE 1.4 ONE-WAY MEASUREMENTS OF THE SPEED OF LIGHT

Observers	Year	V_{eff}
Cole [18]	1976	70 m/s
Krisher et al. [19]	1990	100
Wolf and Petit [20]	1997	0.6

REFERENCES

1. Galileo Galilei, *Dialogue of the Two Principal Systems of the World* (1632). Translated by Thomas Salusbury, London (1661).
2. Isaac Newton, *Principia*, vol. I, (1687).
3. A. A. Michelson, Am. J. Sci. **122**, 120 (1881); A. A. Michelson and E. W. Morley, Am. J. Sci. **134**, 333 (1887).
4. R. S. Shankland, S. W. McCuskey, F. C. Leone, and G. Kuerti, Rev. Mod. Phys. **27**, 167 (1955).
5. L. Essen, Nature **175**, 793 (1955).
6. T. S. Jaseja, A. Javan, J. Murray, and C. H. Townes, Phys. Rev. **133**, 1221 (1964).
7. A. Brillet and J. L. Hall, Phys. Rev. Lett. **42**, 549 (1979).
8. A. A. Michelson and H. G. Gale, Astrophys. J. **61**, 140 (1925).
9. R. J. Kennedy and E. M. Thorndike, Phys. Rev. **42**, 400 (1932).
10. D. Hils and J. L. Hall, Phys. Rev. Lett. **64**, 1697 (1990).
11. J. Müller and M. H. Soffel, Phys. Lett. A **198**, 71 (1995).
12. J. P. Cedarholm, G. F. Bland, and C. H. Townes, Phys. Rev. Lett. **1**, 342 (1958).
13. H. Hay, J. Schiffer, T. Cranshaw, and P. Egelstaff, Phys. Rev. Lett. **4**, 165 (1960).
14. D. C. Champeney, G. R. Isaak, and A. M. Khan, Rev. Mod. Phys. **36**, 469 (1964); Nature **198**, 1186 (1963).
15. K. C. Turner and H. A. Hill, Bull. Am. Phys. Soc. **8**, 28 (1963).
16. G. R. Isaak, Phys. Bull. **21**, 255 (1970).
17. E. Riis, L. A. Andersen, N. Bjerre, O. Poulsen, S. A. Lee, and J. L. Hall, Phys. Rev. Lett. **60**, 81 (1988).
18. T. W. Cole, Mon. Not. R. Astr. Soc. **175**, 93P (1976).
19. T. P. Krisher, L. Maleki, G. F. Lutes, L. E. Primas, R. T. Logan, and J. D. Anderson, Phys. Rev. D **42**, 731 (1990).
20. P. Wolf and G. Petit, Phys. Rev. A **56**, 4405 (1997).

PROBLEMS

1. Suppose we adjust the clocks in the reference frame of the Earth according to local solar time. By means of an example, show that with this synchronization of clocks, Newton's Second Law is not valid in the reference frame of the Earth. Ignore centrifugal and Coriolis effects for the purpose of this question. [*Hint*: Consider an airliner flying from Washington, D.C., to San Francisco. According to *local* time, this flight takes 2 hours. According to local time, how long does the return flight take? Explain why this is a conflict with Newton's Second Law.]

2. Sketch the worldline of a particle that moves with a constant acceleration for $t < 0$ and for $t > 0$ and passes through the origin at $t = 0$.

3. Plot the worldline for an automobile that is initially rest at $x = 0$, begins to accelerate at $t = 0$ and maintains a constant acceleration of 3 m/s^2 until it reaches a final speed of 25 m/s, and then brakes with a constant deceleration of 9 m/s^2 until it is again at rest.

4. Consider a particle bouncing back and forth elastically (with constant speed) between two rigid walls at $x = 0$ and at $x = L$. Draw the worldline for this particle.

5. Sketch the worldline of a planet in an elliptical orbit around the Sun in a three-dimensional spacetime diagram with two space dimensions and one time dimension.

6. Figure 1.8 shows the grid of x',t' coordinates for a spaceship reference frame moving in the positive x direction (that is, $V > 0$). Draw the grid of x',t' coordinates for the case of motion in the negative x direction (that is, $V < 0$).

7. Draw the grids of x,t and x',t' axes, as in Fig. 1.8, and on this grid draw the worldline of a particle that bounces back and forth elastically between two "walls" at rest in the spaceship frame at the points $x' = 0$ and $x' = L$. Note that in the reference frame of the spaceship, this particle has constant speed, with $v'_x = \pm v'$.

8. An alien spaceship traveling toward the Earth at 1.5×10^8 m/s (that is, half the speed of light) launches a probe toward the Earth. In the reference frame of the spaceship, the speed of the probe is 2.5×10^8 m/s. According to the Galilean addition law, what is the speed of the probe relative to the Earth? If astronomers on the Earth are observing the spaceship with a telescope, can they see this probe before it hits them?

9. Galileo thought that the ocean tides result from a combination of the Earth's translational motion and its rotational motion about its axis. Is this consistent with the Galilean Principle of Relativity?

10. A wind is blowing over a lake from the northeast at 20 mi/h. A sailboat is heading east at 8 mi/h. What are the magnitude and the direction of the wind in the reference frame of the sailboat?

11. A tennis ball moving horizontally toward the left at speed v collides with a rigid wall and bounces back elastically. In the reference frame of the ground, the initial speed of the tennis ball is v, and the final speed is also v. What are the initial and final speeds of the tennis ball in a reference frame moving at speed V toward the right with respect to the ground? What is the loss of kinetic energy of the tennis ball in this reference frame? Does this mean that the collision is not elastic and that the Principle of Relativity is violated?

12. In the laboratory reference frame, a particle of mass m is accelerating along the x axis under the influence of a constant force F. According to the work-energy theorem, during a displacement from x_1 to x_2 (in some given time interval), the change of kinetic energy of the particle is

$$\tfrac{1}{2}mv_2^2 - \tfrac{1}{2}mv_1^2 = F \times (x_2 - x_1)$$

During the same time interval, what is the change of kinetic energy $\tfrac{1}{2}mv'^2_2 - \tfrac{1}{2}mv'^2_1$ in the reference frame of a spaceship moving at speed V along the x axis? What is the work done by the force F in the spaceship frame during the same time interval? Is the work-energy theorem valid in the spaceship frame?

13. Consider a reference frame moving with uniform velocity $\mathbf{V}$ through the ether. Show that according to the Galilean addition law, in this reference frame a light signal traveling at an angle θ' with the direction of $\mathbf{V}$ has a speed of

$$c' = \sqrt{c^2 - V^2 \sin^2 \theta'} - V \cos \theta'$$

Verify that this gives the expected result for a parallel light signal ($\theta' = 0$) and for a perpendicular light signal ($\theta' = 90^0$).

14. Consider an ideal gas confined in a container. In the reference frame of the container, the gas particles have a total average kinetic energy $K = 3NkT/2$, where T is the temperature, N is the number of gas particles, and k is Boltzmann's constant. According to Eq. (22), what is the total average kinetic energy K' in a reference frame in which the container has a speed V? What is the value of K' if $T \simeq 0$?

15. In the reference frame of the center of mass of the Sun-Earth system, the Earth is in a circular orbit with constant speed, and the energy and the orbital angular momentum of the Earth are constant (ignore the other planets for the present purposes). But in a reference frame moving in a straight line with uniform velocity $\mathbf{V}$ in the plane of the orbit, the Earth has a cycloidal trajectory (see Fig. 1.19). Is the energy of the motion of the Earth in this reference frame a constant? Is the angular momentum of the Earth calculated about any fixed point in this reference frame a constant? Are the energy and the angular momentum of the Sun-Earth system in this reference frame constant? Is the Principle of Relativity satisfied?

16. Suppose that in the ether frame the Earth has a velocity $\mathbf{V}$ and a ray of light approaches the Earth at an angle θ with respect to the direction of the velocity.
(a) Show that in the reference frame of the Earth, the angle of approach of the ray of light is θ', with

$$\tan \theta' = \frac{\sin \theta}{\cos \theta + V/c}$$

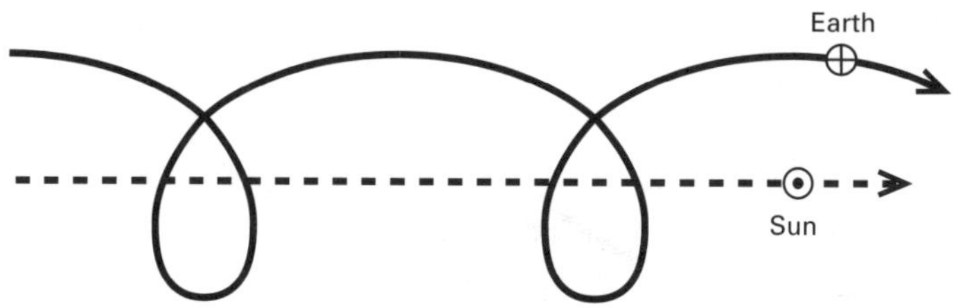

Fig. 1.19 In a reference frame moving with uniform velocity, the Earth has a cycloidal trajectory, that is, a combination of straight-line motion and uniform circular motion.

(b) This formula gives us the aberration of starlight. If $\theta = 90^0$ (for starlight from a star located directly above the orbit of the Earth), what is the deviation of θ' from 90^0?

17. If the Earth drags along a layer of ether with its own velocity **V**, then a ray of light that enters this layer will be dragged along by this layer of ether, and the ray will acquire an extra velocity **V**, additional to whatever velocity it had before entering the layer. Show that this implies that astronomers on Earth would observe no aberration, that is, the angle at which a ray of starlight propagates with respect to the Earth is equal to the angle at which the ray propagates with respect to the external (undragged) ether.

18. In Section 1.6 we calculated the phase difference between the two beams of light in the Michelson-Morley interferometer on the assumption that one arm was

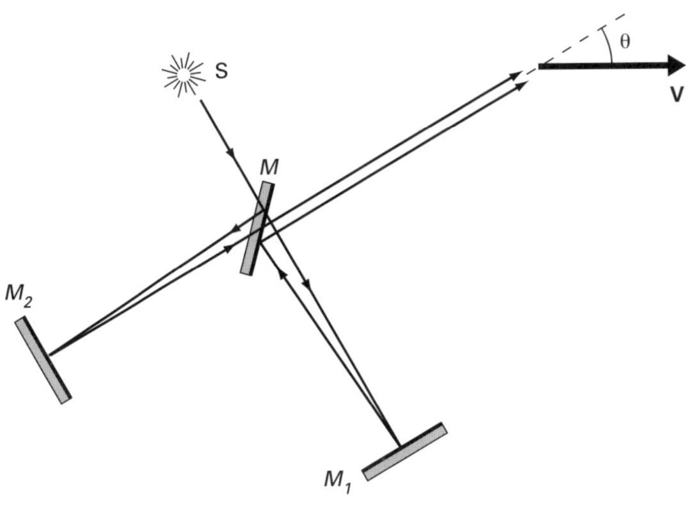

Fig. 1.20 Michelson-Morley interferometer oriented at angle θ with respect to the ether wind.

parallel to the ether wind and the other was perpendicular. Repeat the calculation of the phase difference if one of the arms makes an angle of θ with the ether wind, and the other an angle of $90^0 + \theta$, as in Fig. 1.20. (*Hint*: Use the result of Problem 13, and assume $V \ll c$).

19. (a) Calculate the phase difference between the two beams of light in a Michelson-Morley interferometer with arms of unequal lengths, L_1 and L_2, the first of which is perpendicular to the ether wind, the second parallel to it. Verify that if a Fitzgerald-Lorentz contraction is introduced for the parallel arm, the phase difference becomes

$$\Delta\phi = \frac{2\pi}{\lambda}\frac{2(L_2 - L_1)}{\sqrt{1 - V^2/c^2}} \simeq \frac{2\pi}{\lambda}[2(L_2 - L_1) + (L_2 - L_1)V^2/c^2]$$

(b) Recalculate the phase difference if the arm L_1 is parallel to the ether wind and the arm L_2 perpendicular. Verify that the phase difference is the same as in (a).
(c) The second term in the bracket on the right side of the equation in part (a) gives the dependence of the phase shift on V^2. The interferometer of Kennedy and Thorndike had $L_2 - L_1 = 16$ cm. With $V = 250$ km/s and $\Delta V = 60$ km/s, what is the expected change in the phase difference in six months?

20. Figure 1.21 is a schematic diagram of the experiment Hoek performed in 1868 (this diagram omits some inessential details). A beam of light is split by the half-silvered mirror M into two parallel beams; one of these travels in air, and the other travels through a water-filled tube. The beams are reunited by the half-silvered mirror M', and they interfere constructively or destructively. The phase difference between the beams arises exclusively from the difference of travel time in the parallel segments of length L in air and in water. If the laboratory with the apparatus is moving through the ether at speed V toward the right, the speed of light in air is $c - V$, and the travel time t_1 for light is $L/(c - V)$. Water has an index of refraction $n = 1.33$, and the speed of light in water depends on the details of the ether theory. There are three alternatives:
(a) According to the naive ether theory, the ether is not disturbed by the motion of the water, and the speed of light in water is c/n in the ether frame, *even when the water moves relative to the ether*. If so, the speed of light in the laboratory frame is $c/n - V$.
(b) According to a proposal of Stokes, the water drags the ether along totally. In that case, the speed of light is c/n in the reference frame of the water, and, in the Hoek experiment, it is then also c/n in the laboratory frame.
(c) According to a proposal of Fresnel, the water drags the ether along only partially, so the speed of light in the original (undragged) ether frame is $c/n + V(1 - 1/n^2)$, and the speed of light in the laboratory frame is $c/n + V(1 - 1/n^2) - V = c/n - V/n^2$.
For each of these three alternatives, calculate the travel time t_2 for light through the water-filled tube. Then calculate the phase difference $2\pi c(t_2 - t_1)/\lambda$, where λ is the wavelength in air. Verify that for alternative (c), the phase difference is nearly

independent of the speed V, except for small terms of order V^2/c^2. Thus, in Fresnel's version of the ether theory, the Hoek experiment cannot detect the ether wind. (In the experiment as performed by Hoek, the situation was even worse: Hoek compared the phases for a beam traveling clockwise around the loop in Fig. 1.21 and a beam traveling counterclockwise. This means that he had to take the difference between the phase calculated for V and the phase for $-V$; in this difference, the only terms that survive are of order V^3/c^3).

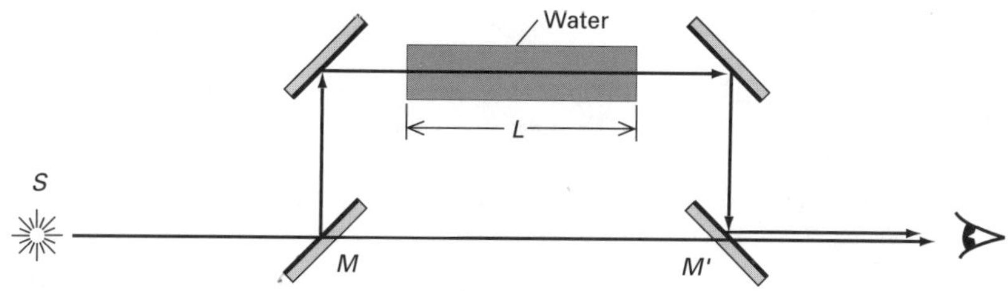

Fig. 1.21 Schematic diagram of the experiment of Hoek.

21. Suppose that a binary star system consists of two stars orbiting about each other in circular orbits with a period of 104 days. The orbital speed of one of the stars is 25 km/s. Suppose the Earth is located in the orbital plane of this system, at a large distance L. According to the ballistic theory of light, the speed of the light emitted by the star is c + 25 km/s when the star is at one extreme of its orbit and approaching the Earth, and it is c - 25 km/s when the star is at the opposite extreme of the orbit and receding from the Earth. If L is sufficiently large, astronomers on the Earth will see the star simultaneously at both extremes of its orbit, that is, they will see two images of this one star "in opposition." What is the critical distance L?

2

Einstein's Theory and the Lorentz Transformation

GOD IS CUNNING, BUT HE IS NOT MALICIOUS.

Albert Einstein, at a reception in Princeton, 1921

I HAVE HAD SECOND THOUGHTS. MAYBE GOD IS MALICIOUS AFTER ALL.

Albert Einstein to Peter Bergmann, ca. 1940

Einstein formulated his theory of Special Relativity in 1905. The null result of the Michelson-Morley experiment was known to Einstein at the time, but this experiment played only a small role in his thinking. Einstein was much more concerned with Maxwell's theory and how transformations of reference frames affect Maxwell's equations. Einstein was fascinated by the mathematical structure of Maxwell's equations, and he was firmly convinced that these equations must be valid in all reference frames. As he described in his autobiographical notes, the conflict between Maxwell's equations and the Galilean transformation equations led him to the fundamental principle of his theory of relativity:

> After ten years of reflection such a principle resulted from a paradox upon which I had already hit at the age of sixteen: If I pursue a beam of light with the velocity c (velocity of light in vacuum), I should observe such a beam as an electromagnetic field constant in time, periodic in space. However, there seems to exist no such thing, neither on the basis of experience, nor according to Maxwell's equations . . . [1]

This contradiction left Einstein with two choices: either Maxwell's equations were wrong in all reference frames except one (the ether frame), or else the Galilean transformation and the Newtonian ideas about spacetime were wrong. Faced with this choice, Einstein decided to retain Maxwell's equations and to discard the Galilean transformation and Newtonian spacetime. He then constructed a new geometry for spacetime based on the requirement that the speed of light must be the same in all inertial reference frames. This led him to a profound revision of our concepts of space and time and to a corresponding profound revision of the laws of mechanics.

2.1 EINSTEIN'S PRINCIPLE OF RELATIVITY

In Chapter 1, we saw that Newtonian mechanics obeys a principle of relativity, but Maxwell's equations do not obey such a principle, at least not according to the ether theory and the Galilean transformation. Einstein laid the foundations of his theory by postulating a **Principle of Relativity** for all the laws of physics, that is, for mechanics, and for Maxwell's equations, and for all other laws of physics:

All the laws of physics, and all the numerical constants that appear in them, are exactly the same in all inertial reference frames.

He supplemented this principle by a separate principle for the propagation of light:*

The speed of light (in vacuum) is the same in all inertial reference frames; it always has the value $c = 3.0 \times 10^8$ m/s.

Here, as in Chapter 1, the value for the speed of light has been rounded off; the exact value is $c = 2.99792458 \times 10^8$ m/s.

Using the terminology introduced in Chapter 1, we can say that according to Einstein's Principle of Relativity, all the laws of physics are invariant under transformations from one inertial reference frame to another, and that according to the principle for the propagation of light, the speed of light is invariant. Both principles (and their consequences) have been subjected to rigorous experimental tests, and they have been confirmed with high precision. We reviewed the experimental evidence for the invariance of the speed of light in the preceding chapter, and we will describe the evidence for various consequences of the Principle of Relativity in the next chapters.

The invariance of the speed of light obviously conflicts with the Galilean addition law for velocities. We therefore have to discard this law, and we also have to discard the Galilean transformation equations of the space and time coordinates on which this law is based. In Sec-

* Although Einstein stated the invariance of the speed of light as a separate principle, the two principles are not entirely independent. If Maxwell's equations are valid laws of physics, then we can deduce the value of the speed of light from Maxwell's equations, and then the first principle implies the second. Hence the second principle is to be regarded as "insurance" in case Maxwell's equations are not valid laws of physics.

tion 2.3 we will formulate a new set of transformation equations, the Lorentz transformation equations, which replace the Galilean transformation equations.

The invariance of the speed of light also forces us to abandon some of the intuitive notions of space and time suggested by our everyday experience. The fact that a light signal always has the same speed of 3.0×10^8 m/s relative to us, regardless how fast we try to move toward or away from this signal in a fast aircraft or a fast spaceship (or a fast planet), is in conflict with our intuition. This strange behavior of light signals is possible only because of a strange behavior of length and time in relativistic physics. We have to abandon our intuitive Newtonian concepts of absolute length and absolute time. As we will see in this chapter, in relativistic physics, lengths and times are relative; they both depend on the reference frame in which they are measured, and they suffer contractions and dilations when we change the reference frame.

2.2 THE SYNCHRONIZATION OF CLOCKS AND THE RELATIVITY OF SIMULTANEITY

Einstein was acutely aware that his principles required a profound modification of our concepts of space and time. To discover the required modification of the properties of space and time, Einstein examined the implications of the invariance of the speed of light, and from this he deduced a new set of transformation equations of the space and time coordinates. To obtain these transformation equations, we must begin with a careful construction of the inertial reference frames and a careful synchronization of their clocks.

As described in Section 1.1, a reference frame is a lattice of measuring rods with a set of synchronized clocks. The construction of the lattice of measuring rods used for the determination of the x,y,z coordinates poses no special problems--it can be performed in relativity in the same way as in Newtonian physics. However, the synchronization of the clocks used for the determination of the time coordinate t in this reference frame requires some extra attention. In Newtonian physics, the synchronization of two clocks at different positions can be achieved very simply by transporting a third clock from one position to the other and making comparisons. The transported clock must be designed so that its rate is not affected by the accelerations it suffers during the motion, but this poses no serious problem (for instance, a good wristwatch is fairly resistant to accelerations, and we can keep errors in synchronization under control by moving the watch slowly and carefully). In relativistic physics, synchronization is a more delicate matter, because, as we will see, the rate of a clock depends on its speed. To avoid trouble, we can try to move the clock slowly. But if we move a clock slowly in, say, the laboratory frame, its speed will still be large in the spaceship frame, and its rate in that reference frame will be altered. Thus, synchronization in the spaceship frame requires a new transported clock that moves slowly in *that* frame. Synchronization by means of clock transport therefore does not directly tell us how the synchronization in one reference frame is related to that in another reference frame, and we cannot assert, as in the Newtonian case, that synchronization is absolute and that time is absolute.

Einstein proposed an alternative procedure for synchronization, by means of light signals. This will permit us to discover quite directly how the synchronizations in different reference frames are related. But first we want to use light signals to achieve synchronization of two clocks at different positions in a given reference frame. For this purpose, we simply place a light source at a point exactly midway between the two clocks and send out a flash of light toward both clocks. The two clocks are synchronized if both indicate exactly the same time when the light signals reach them (see Fig. 2.1). Obviously, this synchronization procedure hinges on the invariance of the speed of light. If the speed of light were not an invariant quantity, but depended on the reference frame and on the direction of propagation (say, faster toward the right in Fig. 2.1 and slower toward the left), then we could not achieve synchronization by means of a flash emitted from the midpoint.

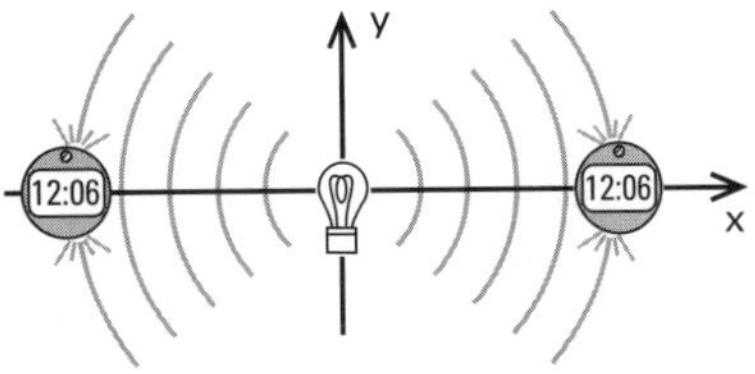

Fig. 2.1 A flash of light is emitted by a source at the midpoint between two clocks. Light signals (indicated by the progressing wavefronts) travel toward the left and the right.

One immediate consequence of this synchronization procedure by light signals is that synchronization is relative, that is, when the clocks in the laboratory frame are synchronized with each other, and the clocks in the spaceship frame are synchronized with each other, the clocks of the laboratory frame are not synchronized with those of the spaceship frame. We can understand this by examining the light signals from the points of view of the laboratory frame and the spaceship frame. In the laboratory frame, the two light signals are emitted from the midpoint, they travel toward the right and the left at equal speeds, and they reach the laboratory clocks simultaneously (see Fig. 2.2a). Now assume that the midpoint of the spaceship coincides with the light source at the instant the flash is emitted. Then, in the spaceship frame, the two light signals can be regarded as emitted from the midpoint of the spaceship, they travel toward the right and the left at equal speeds (as guaranteed by invariance of the speed of light relative to the spaceship), and they reach the spaceship clocks simultaneously (see Fig. 2.2b). Thus, the clocks in the laboratory frame are synchronized with each other, and the clocks in the spaceship frame are synchronized with each other. However, as judged by observers in the laboratory frame, the spaceship clock on the right is moving away from the light source, whereas the spaceship clock on the left is moving toward the light source, and consequently the left clock will intercept the leftward light signal *before* the right clock intercepts the rightward light signal (see Fig. 2.2c). This means that as judged by observers in the laboratory frame, the spaceship clocks are *not* synchronized.

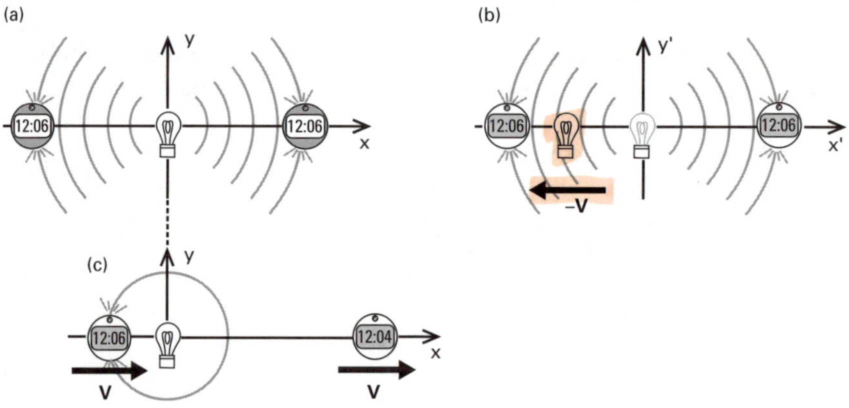

Fig. 2.2 (a) In the laboratory frame, a light source at the midpoint emits a flash of light. The light signals reach the two laboratory clocks simultaneously. (b) In the spaceship frame, the light signals reach the two spaceship clocks simultaneously. Note that the light source was at the midpoint (faint outline of light bulb) when it emitted the light, although it has now moved to the left (solid outline) relative to the spaceship. (c) In the laboratory frame, the left spaceship clock intercepts the leftward signal before the right spaceship clock intercepts the rightward signal.

The converse is also true: As judged by the observers in the spaceship frame, the clock on the left of the laboratory (or the clock at the leading end of the laboratory in its motion relative to the spaceship) indicates a later time than the clock on the right (at the trailing end of the laboratory). In general, whenever one reference frame is moving relative to another, the clocks at the leading end of the moving reference frame are late as judged by the clocks of the other reference frame. To remember this, use the mnemonic *Leading Late*.

It is instructive to examine these laboratory and spaceship synchronization procedures by means of spacetime diagrams showing the worldlines of the clocks and of the light signals used in their synchronization. In relativistic physics, it is customary to use a spacetime diagram with axes x and ct (rather than x and t, as in Chapter 1). With this choice of axes, the worldline of a light signal has a slope of 1, and it makes an angle of 45° with the ct axis and also with the x axis. Figure 2.3 describes the synchronization procedure in the laboratory frame. It shows the worldlines of two clocks at rest in the laboratory frame, and it shows the worldlines of the light signals emitted from the midpoint toward these two clocks. The worldlines of the light signals slope upward at 45° toward the left and the right. These light signals reach the two clocks at the spacetime points P and at Q, at which points the two clocks must indicate equal times. In Fig. 2.3, the dashed line is an equal-time line, or the locus of spacetime points simultaneous with P and Q. This line is parallel to the x axis, as expected.

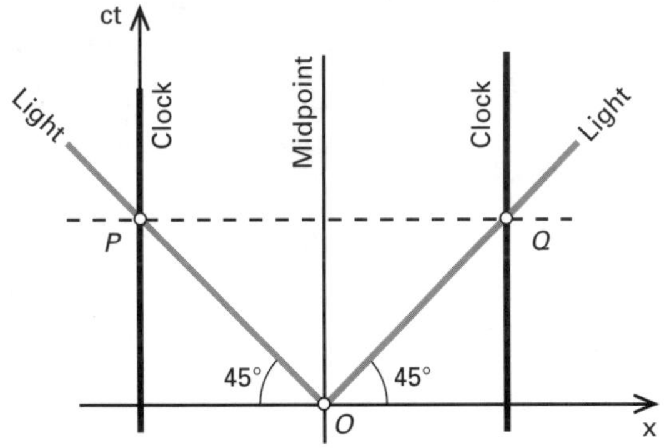

Fig. 2.3 Worldlines of two clocks at rest in the laboratory frame and worldlines OP and OQ of the two light signals toward the left and the right. The dashed line PQ is an equal-time line in the laboratory frame.

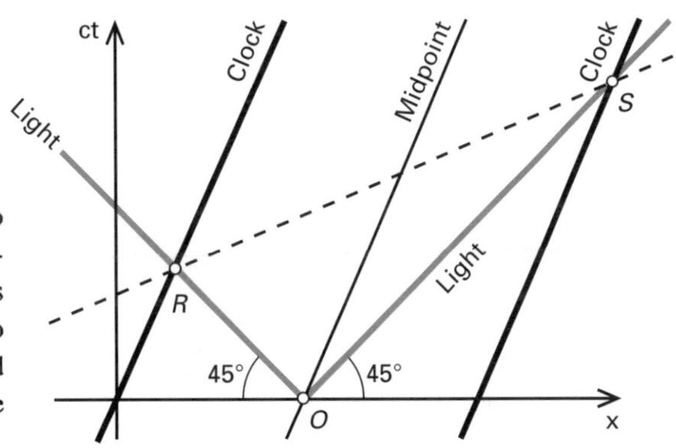

Fig. 2.4 Worldlines of two clocks at rest in the spaceship frame and worldlines OR and OS of the two light signals. The dashed line RS is an equal-time line in the spaceship frame.

Figure 2.4 describes the synchronization procedure in the spaceship frame. In the spacetime diagram, the worldline of the center of the spaceship is the line $x = Vt$. Expressed in terms of the coordinate ct, this is $x = (V/c)ct$, which is a line of slope V/c relative to the ct axis. Each spaceship clock moves at the same speed as the center of the spaceship; thus, all their worldlines have the same slope. Figure 2.4 shows the worldlines of two such clocks at rest in the spaceship frame, and it shows the worldlines of the light signals emitted from the midpoint toward these two clocks. Again the worldlines of these light signals slant upward at 45°. The light signals reach the two clocks at R and S, at which points the spaceship clocks must indicate equal times. But the points R and S clearly correspond to *unequal* laboratory times! Thus, although the spaceship clocks are synchronized in their own reference frame, they are not synchronized in the laboratory reference frame. As judged by the laboratory clocks, the clock on the right of the spaceship (or the clock at the leading end of the spaceship) indicates a later time than the clock on the left (at the trailing end of the spaceship). This discrepancy in the synchronizations corresponds to the fact that, as judged in the laboratory, the clock on the right is moving away from the light signal used for synchronization, and the clock on the left is moving toward the light signal. Hence the signal traveling toward the left reaches the left clock more quickly.

In the spacetime diagram, the x axis is the equal-time line $t = 0$, which is the locus of points that are simultaneous with the origin in the laboratory frame. Likewise, the x' axis is the equal-time line $t' = 0$, which is the locus of points that are simultaneous with the origin in the spaceship frame. In Figure 2.4, the dashed line RS is an equal-time line in the spaceship frame--it is the locus of points that are simultaneous with R and S in the spaceship frame. Since all the equal-time lines in the spaceship frame must be parallel to each other, the x' axis must be parallel to the line RS. Thus the x' axis is slanted upward relative to the x axis (see Fig. 2.5). This slant is a graphical expression of the relativity of synchronization.

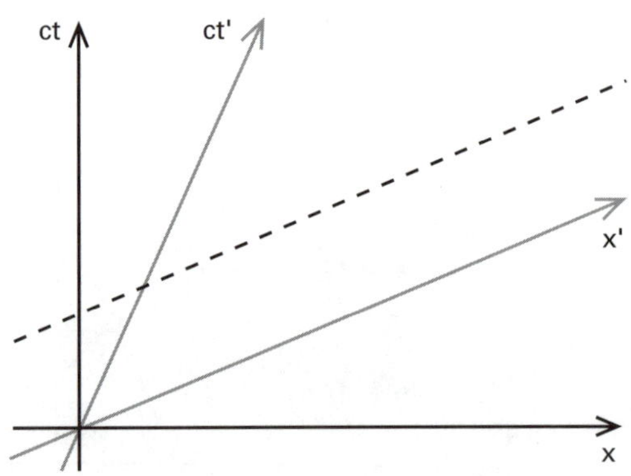

Fig. 2.5 The x' axis is parallel to the equal-time line.

The angle between the x' axis and the x axis is equal to the angle between the ct' axis and the ct axis. This can be easily seen by thinking of the worldlines of the light signals used for the synchronization as forming two sides of a rectangle (see Fig. 2.6); the symmetry of this rectangle and its inscribed triangles reveals the equality of the angles. Accordingly, both the ct' axis and the x' axis are slanted, with equal slant angles (see Fig. 2.7). The angle between the ct' axis and the ct axis is $\phi = \tan^{-1} V/c$, and the angle between the x' axis and the x axis

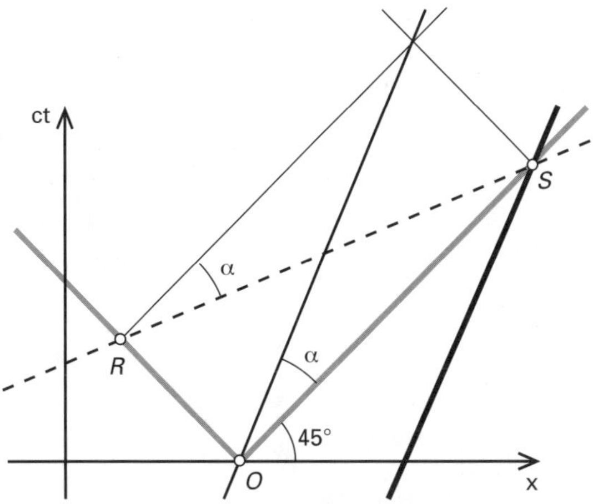

Fig. 2.6 The worldlines *OR* and *OS* of the light signals from Fig. 2.4 form the two lower sides of a rectangle. The other two sides are constructed by drawing parallels. The two angles α are equal, by the symmetry of the rectangle. The dashed diagonal makes an angle of $45^0 - \alpha$ with the x axis. The other diagonal makes an angle of $45^0 + \alpha$ with x axis and, hence, an angle of $90^0 - 45^0 - \alpha = 45^0 - \alpha$ with the ct axis. The x' axis is parallel to the dashed diagonal, and the ct' axis is parallel to the other diagonal. Hence the angle between the x' and x axes is $45^0 - \alpha$, and the angle between the ct' and ct axes is also $45^0 - \alpha$, that is, the slant angles of the x' and ct' axes are equal.

is also $\phi = \tan^{-1} V/c$. The coordinate transformation from the laboratory frame to the spaceship frame is a transformation from the rectangular axes ct,x to the (symmetrically) slanted axes ct',x'. The coordinates x,ct and the coordinates x',ct' of any event P can be found by drawing appropriate intercepts toward the axes; the intercept lines toward the x',ct' axes are slanted, parallel to the axes (see Fig. 2.7). Note that Fig. 2.7 gives only a qualitative determination of the coordinates. For a quantitative determination, we need to know the calibration of the new axes in relation to the old, that is, we need to know the location of the unit points on the axes. We cannot establish this calibration by examining the synchronization of clocks; we need to

Fig. 2.7 The angle between the ct' axis and the ct axis is $\phi = \tan^{-1} V/c$, and the angle between the x' axis and the x axis is also $\phi = \tan^{-1} V/c$. The coordinates x',ct' of a spacetime point P are found by drawing slanted intercepts.

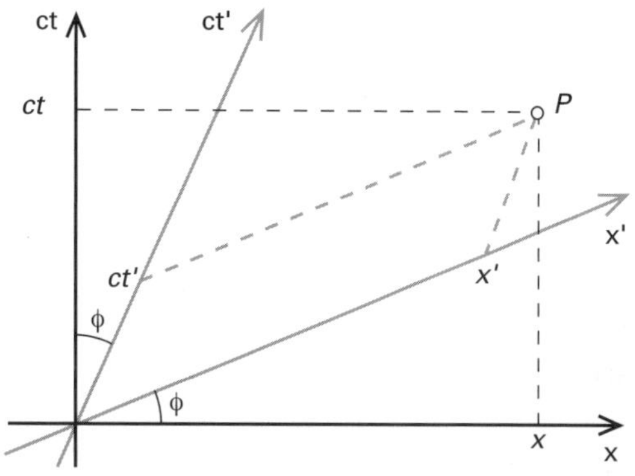

provide some extra information. We will do this in the next section, where we deal with the full details of the coordinate transformation.

The relativity of the synchronization of clocks has as a direct consequence the relativity of the simultaneity of events measured by these clocks. For a concrete example, suppose that an automobile accident occurs at exactly 10:00 A.M. Eastern Standard Time in Philadelphia and that another automobile accident occurs at exactly 10:00 A.M. in Boston. These accidents are simultaneous in the reference frame of the Earth. But in the reference frame of a fast spaceship traveling in the direction from Phildelphia to Boston, the accidents are not simultaneous. Relative to the spaceship, the Philadelphia clock is at the leading end of the Earth reference frame, and it is late, so the Philadelphia accident occurs later than the Boston accident.

Thus, two events that are simultaneous in one reference frame can become nonsimultaneous in another. And even worse, two events that occur with one time order in one reference frame can occur with the reverse time order in another reference frame. Such a reversal of the time order is possible whenever the slope of the line connecting the two events in the spacetime diagram is less than 1, which means the spatial separation Δx between the events is larger than the temporal separation Δt. Figure 2.8 shows two such events O and P. In the laboratory reference frame, O occurs before P. But in the reference frame of a fast spaceship, with x',ct' axes as shown in Fig. 2.8, P occurs before O. Obviously, this reversal of the time order hinges on finding a spaceship reference frame such that the upward slant of the x' axis exceeds the upward slant of the line OP. This is possible whenever the upward slant of the line OP is less than 45°, that is, whenever the slope is less than 1.

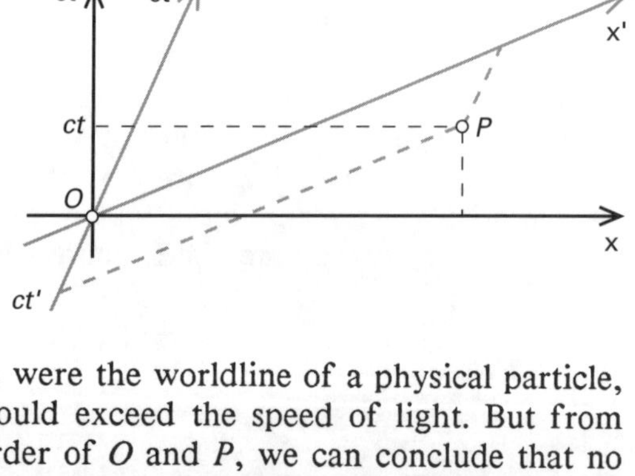

Fig. 2.8 The ct coordinate of the spacetime point P is positive, but the ct' coordinate is negative.

If the line OP in Fig. 2.8 were the worldline of a physical particle, the speed of that particle would exceed the speed of light. But from the ambiguity of the time order of O and P, we can conclude that no physical particle may travel along such a worldline. Our basic notions of causality require that if the particle departs from the point O and later arrives at a point P, this time order must be the same in all reference frames. (For instance, imagine that the "particle" is a kitten born at O; it evidently cannot arrive at P before it is born at O.) Consequently, a physical particle cannot travel along the worldline OP of

Fig. 2.8, or, more generally, it cannot travel along any worldline of slope less than 1. That is, a physical particle cannot have a speed exceeding the speed of light.

More generally, a physical signal of any kind cannot have a speed exceeding the speed of light. A direct consequence of this limitation is that an absolutely rigid body cannot exist, because such a rigid body could be used to transmit signals with infinite speed. For instance, a sudden push exerted against one end of an absolutely rigid rod would cause an immediate displacement of the other end of the rod, which would constitute a signal with infinite speed. Physical rods made of solid materials are always somewhat elastic. They are stiff but not absolutely rigid, and the speed of a compressional signal propagating along the rod depends on the speed of sound in the solid material; this speed is always much lower than the speed of light.

For two events connected by a line of slope equal to or more than 1, the time order is the same in all reference frames. Thus, the time order is absolute for any two events that are connected by a light signal (with a worldline of slope 1) or by a signal consisting of a messenger particle (that is, a particle that moves from one event to the other with a worldline of slope larger than 1). This is consistent with our notions of causality, since the transmission of the signal from the first event and the reception of this signal at the second event are causally related, and an ambiguity in their time order would be nonsensical. For instance, it would be nonsensical that a letter be received at its destination before the sender mails it at the place of origin.

2.3 THE LORENTZ TRANSFORMATION

In Einstein's relativistic physics, the relationship between the space and time coordinates in two inertial reference frames is given by the **Lorentz transformation**. This transformation replaces the Galilean transformation of Newtonian physics. Whereas in the Newtonian case we relied on the absolute character of length and of time to derive the transformation, in the relativistic case we have to rely on the invariance of the speed of light. In the preceding section we examined the implications of this invariance of the speed of light for the synchronization of clocks in two inertial reference frames, and we saw that the time coordinate is not absolute, that is, in general $t' \neq t$. We saw that the transformation from the x,ct coordinates to the x',ct' coordinates is a transformation to slanted axes--both the x' axis and the ct' axis are slanted. In this section we will deal with the details of the Lorentz transformation; in particular, we will deal with the calibration of the slanted axes. To discover these details, we take as a starting point what we learned about the slant of the axes in the preceding section and what we know about the calibration of the axes in the Galilean case.

From Section 1.3, we recall that the Galilean transformation equations are

$$t' = t \tag{1}$$

$$x' = x - Vt \tag{2}$$

$$y' = y \tag{3}$$

$$z' = z \tag{4}$$

Here we have assumed that the second reference frame moves with velocity V in the direction of the x axis of the first reference frame and that the origins of the two reference frames coincide at the time $t = t' = 0$. This transformation leaves distances (at a given time) invariant, and it leaves time intervals invariant. We know that geometrically the Galilean transformation is a transformation to a slanted t' axis. The x' axis coincides with the x axis, but the t' axis lies along the slanted line $x = Vt$ (see Fig. 2.9). This means that the t' axis is the worldline of the origin ($x' = 0$) of the moving reference frame, or the worldline of the center of the spaceship, as it should be. Finally, we note the geometrical procedure for the calibration of the t' axis: To find the unit point on the t' axis, we start at the unit point on the t axis and draw a horizontal line, which intercepts the t' axis at its unit point (see Fig. 2.9).

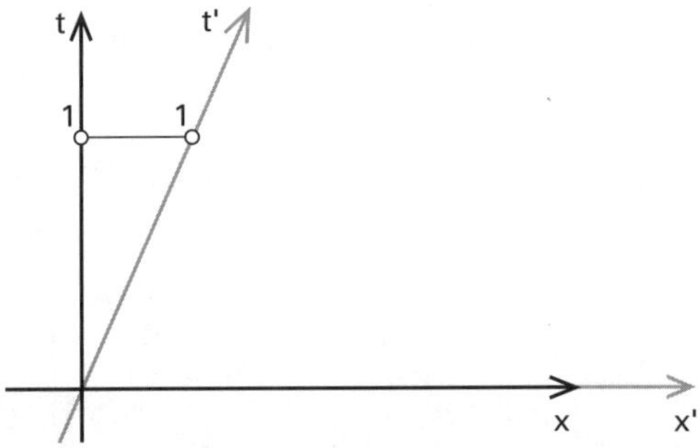

Fig. 2.9 Arrangement of axes for the Galilean transformation. The t' axis is slanted, but the x' axis is not. The t' axis is calibrated by drawing a horizontal line from $t = 1$, which intercepts the t' axis at $t' = 1$.

To derive the Lorentz transformation, we begin by drawing the ct' and x' axes in the spacetime diagram. The ct' axis is simply the worldline of the origin ($x' = 0$) of the spaceship frame. Thus, as in the Newtonian case, it is simply the line $x = Vt$, or $x = (V/c)ct$, or a line of slope V/c relative to the ct axis. The drastically new feature of the relativistic case is the location of the x' axis; in contrast to the Newtonian case, it does not coincide with the x axis. As we saw in the preceding section, the x' axis is slanted relative to the x axis, with the same slant angle ϕ as the ct' axis relative to the ct axis. Figure 2.10 shows these slanted x' and ct' axes in the spacetime diagram. Such a diagram, with two sets of axes, is called a **Minkowski diagram** after the mathematician H. Minkowski, a contemporary of Einstein.

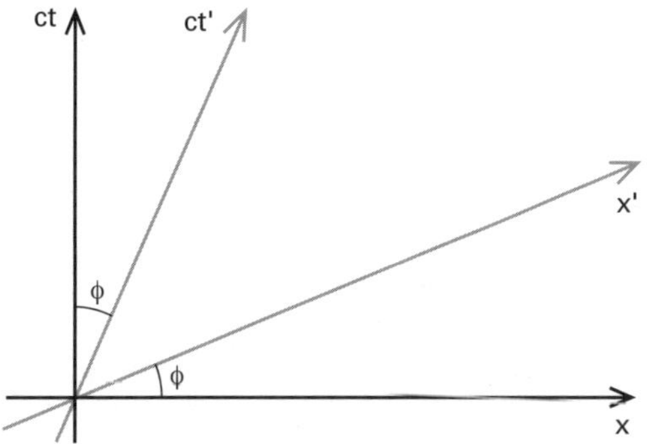

Fig. 2.10 Arrangement of axes for the relativistic coordinate transformation, or Lorentz transformation. This diagram, with slanted x' and ct' axes, is called a Minkowski diagram.

The symmetric arrangement of the slanted axes implies that the two sets of coordinates are related by symmetric equations of the form

$$x' \propto x - \frac{V}{c}\,ct \tag{5}$$

$$ct' \propto ct - \frac{V}{c}\,x \tag{6}$$

The first of these equations is essentially the same as Eq. (2). It expresses the slant of the t' axis and the corresponding slant of the intercept line that determines the x' coordinate. The second equation expresses the slant of the x' axis. This second equation follows from the first by symmetry, that is, by exchange of the space and time coordinates. This symmetry ensures that the angles between the x,x' and ct,ct' axes are equal. The construction of these slanted axes incorporates the relativistic invariance of the speed of light and the relativity of synchronization of clocks.

Note that Eqs. (5) and (6) are stated as proportionalities rather than as equalities. The reason is that the slant angles ϕ in Fig. 2.10 fix the directions of the ct' and x' axes, but not the calibration of these axes, or the locations of the unit points on the axes. If we include the unknown factor of proportionality in our equations, we can write them as

$$x' = \gamma\left(x - \frac{V}{c}\,ct\right) \tag{7}$$

$$ct' = \gamma\left(ct - \frac{V}{c}\,x\right) \tag{8}$$

Here, the same factor of proportionality γ appears in both equations because the worldline of a light signal emitted from the origin, with x

$= ct$ in the laboratory frame, must become $x' = ct'$ in the spaceship frame. This requires the same factor γ in both Eqs. (7) and (8).

The factor γ is some function of the velocity V, which is determined by the calibration of the axes. To find the correct calibration, let us construct the Lorentz transformation for a finite velocity V by means of a sequence of transformations with infinitesimal velocity increments. Figure 2.11 shows the axes x',ct' and x'',ct'' for two sequential transformations with small velocity increments. For such small velocity increments, we can perform the calibration of the time axes by the same geometrical procedure as in the Galilean case: To find the unit point on the ct' axis, we start at the unit point on the ct axis and draw a line parallel to the x axis. And to find the unit point on the ct'' axis, we start at the unit point on the ct' axis and draw a line parallel to the x' axis, etc. (see Fig. 2.11). By this step-by-step procedure with small velocity increments, we can approximately calibrate a sequence of axes ct', ct'', ct''', etc., and we can ultimately find the unit point on the time axis for a Lorentz transformation of a large velocity.

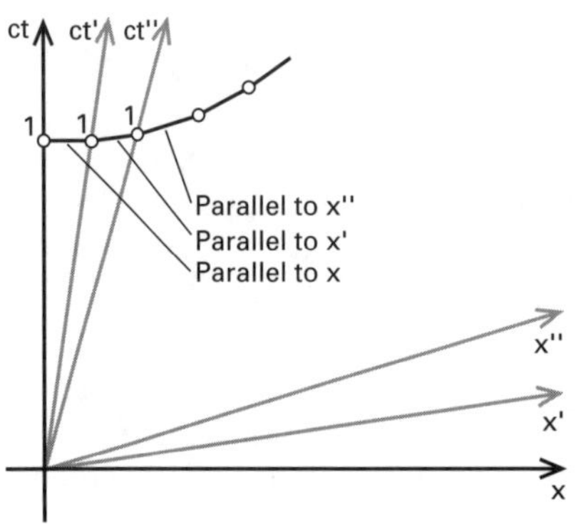

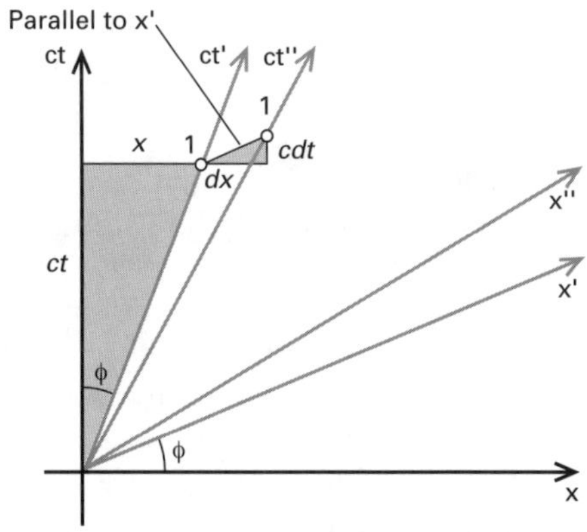

Fig. 2.11 (Left) Axes x',ct' and x'',ct'' and x''',ct''' for a sequence of Lorentz transformations with small velocity increments. The unit points on the ct' axis, the ct'' axis, and the ct''' axis are obtained by drawing successive line segments parallel to the x axis, the x' axis, and the x'' axis.

Fig. 2.12 (Right) Axes x',ct' for a Lorentz transformation of some given velocity, and axes x'',ct'' for a slightly larger velocity. The shaded triangles are similar.

In the limiting case of infinitesimal velocity increments, the polygonal line connecting the unit points on the ct, ct', ct'', ct''', etc. axes becomes a smooth curve which determines the exact calibration of the time axis, for any arbitrary velocity. This calibration curve connecting the unit points on the time axes is a hyperbola. To prove this, consider the axes x',ct' that correspond to a Lorentz transformation of some given velocity, and perform an infinitesimal increment of this velocity. Figure 2.12 shows the axes x',ct' and the axes x'',ct'' that correspond to the given velocity plus an infinitesimal increment. The unit point on the ct'' axis is obtained from the unit point on the ct' axis by a horizontal displacement dx and a vertical displacement cdt. The shaded right triangles are similar, because the acute angle in each equals the angle ϕ between the x' and x axes. The ratios of their opposite and adjacent sides are then equal:

$$\frac{dx}{cdt} = \frac{ct}{x}$$

Hence $c^2tdt - xdx = 0$, and by integration we find that the equation of the calibration curve is $c^2t^2 - x^2 = \text{constant}$. Since $x = 0$ corresponds to $ct = 1$, we see that constant = 1, and thus the equation of the calibration curve is

$$c^2t^2 - x^2 = 1 \tag{9}$$

This is the equation of a hyperbola, with apex at the unit point on the ct axis and asymptotes along the 45^0 lines (see Fig. 2.13).

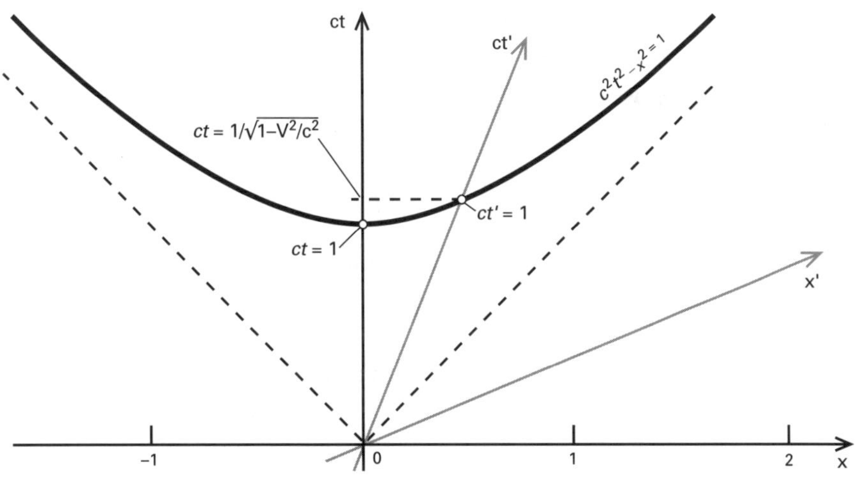

Fig. 2.13 The calibration curve connects the unit points $ct = 1$ and $ct' = 1$ on the time axes. The calibration curve is a hyperbola $c^2t^2 - x^2 = 1$ with apex at $ct = 1$, and asymptotes at 45^0. A horizontal line drawn from the unit point $ct' = 1$ intercepts the ct axis at $ct = 1/\sqrt{1 - V^2/c^2}$.

The calibration curve determines the unit point on the ct' axis and therefore permits us to calculate the ct coordinate of this unit point. This unit point is at the intersection of the ct' axis and the calibration curve. In terms of the ct,x coordinates, the ct' axis has the equation x

$= Vt$. The intersection of the line $x = Vt$ and the curve $ct^2 - x^2 = 1$ therefore corresponds to $ct^2 - V^2t^2 = 1$, or

$$ct = 1/\sqrt{1 - V^2/c^2} \tag{10}$$

This value is marked in Fig. 2.13. Note that when we are drawing Minkowski diagrams, Eq. (10) provides us with a convenient alternative for calibrating the ct' axis: we simply draw a horizontal line from the point $ct = 1/\sqrt{1 - V^2/c^2}$, which intercepts the ct' axis at its unit point.

From the known calibration of the ct' axis we can determine the constant of proportionality γ in Eqs. (7) and (8). To evaluate this constant, let us compare the value (10) of the ct coordinate of the unit point on the ct' axis with the value implied by Eq. (8). When we substitute $x = Vt$ and $ct' = 1$ in Eq. (8), we find

$$1 = \gamma\left(ct - \frac{V^2}{c}t\right)$$

which implies

$$ct = \frac{1}{\gamma(1 - V^2/c^2)}$$

Comparing this with Eq. (10), we see that

$$\gamma = \frac{1}{\sqrt{1 - V^2/c^2}} \tag{11}$$

With this expression for γ, our transformation equations (7) and (8) become

$$x' = \frac{1}{\sqrt{1 - V^2/c^2}}\left(x - \frac{V}{c}\,ct\right) \tag{12}$$

$$ct' = \frac{1}{\sqrt{1 - V^2/c^2}}\left(ct - \frac{V}{c}\,x\right) \tag{13}$$

or

$$\boxed{x' = \frac{x - Vt}{\sqrt{1 - V^2/c^2}}} \tag{14}$$

$$\boxed{t' = \frac{t - Vx/c^2}{\sqrt{1 - V^2/c^2}}} \tag{15}$$

These are the **Lorentz transformation equations.** By their construction, these equations incorporate the relativistic invariance of the speed of light, as required by the Principle of Relativity. These equations were first discovered by W. Voigt (1887) and then rediscovered by J. Larmor (1900) and by H. A. Lorentz (1903). But Lorentz and his predecessors thought that these equations were merely a mathematical trick for manipulating solutions of Maxwell's equations, and they had no idea of their deeper meaning for relativity.

In all the above equations, the quantity c is the speed of light in vacuum, and this is so even in regions that are filled with some dielectric material. For instance, in the atmosphere of the Earth, the actual speed of light is less than c, and this actual speed is not invariant under transformations of the reference frame; nevertheless, the Lorentz transformation equations still apply in the form given by Eqs. (14) and (15), since these equations express a fundamental property of the spacetime geometry which is unaffected by the presence of air. Another way to understand this persistent validity of the Lorentz transformation equations in regions filled with dielectric material is by recalling that the dielectric constant depends on the frequency of the light and that for the limiting case of extremely high frequency (gamma rays), the dielectric constant approaches 1 in every kind of dielectric material. Einstein's principle for the propagation of light is therefore always valid for light of extremely high frequency, and for such light all the arguments leading to Eqs. (14) and (15) remain valid in regions filled with dielectric. We could even discard Einstein's principle for the propagation of light and replace it by a postulate of the existence of an invariant maximum signal speed c, without specifying the nature of the signal [2, 3]. It is only because of historical reasons that the invariance of the speed of light has become so closely associated with relativity and the Lorentz transformations.

The inverse Lorentz transformation equations are

$$x = \frac{1}{\sqrt{1 - V^2/c^2}}(x' + Vt') \tag{16}$$

$$t = \frac{1}{\sqrt{1 - V^2/c^2}}(t' + Vx'/c^2) \tag{17}$$

These equations can be obtained by solving Eqs. (14) and (15) for x,t. Alternatively, they can be obtained by exchanging the primed and unprimed coordinates in Eqs. (14) and (15) and reversing the velocity from V to $-V$. This exchange is in accord with the relationship between the laboratory and the spaceship reference frames: if the velocity of the spaceship relative to the laboratory is positive, then the velocity of the laboratory relative to the spaceship is negative.

In the derivation of these transformation equations, we have focused on just one space coordinate, the x coordinate, which is the coordinate measured parallel to the direction of the relative motion of the reference frames. The Lorentz transformation equations for the coordinates y and z, which are perpendicular to the direction of motion, are the same as the Galilean transformation equations for these coordinates,

$$y' = y \tag{18}$$

$$z' = z \tag{19}$$

These equations assert that lengths transverse to the direction of the velocity remain unchanged by the transformation of reference frame. A simple argument for this is the following: Imagine that a piece of pipe is at rest in the laboratory frame, and another identical piece of pipe is at rest in the spaceship frame (see Fig. 2.14). If the motion of the spaceship pipe relative to the laboratory were to cause, say, a contraction of the transverse diameter of the pipe, then--by the principle of relativity--the motion of the laboratory pipe relative to the spaceship would have to cause a contraction of the laboratory pipe. These contraction effects are mutually contradictory since in one case the second pipe would fit inside the first pipe, and in the other case it would fit outside. The only logical alternative is that neither pipe contracts; that is, transverse lengths are unchanged, as asserted in Eqs. (18) and (19).

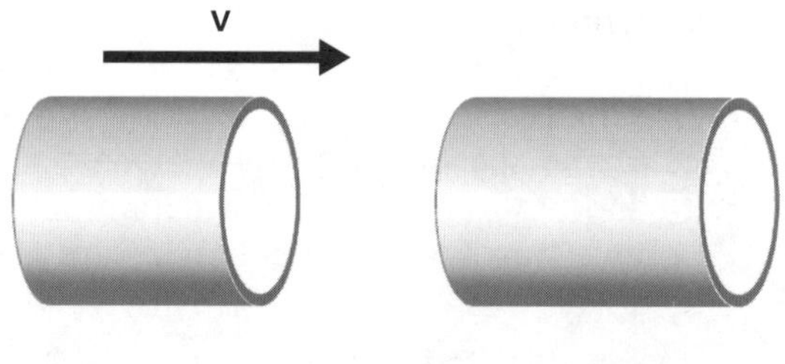

Fig. 2.14 Two equal pieces of pipe, one at rest and one moving at velocity **V**.

Given any laboratory coordinates x,ct, we can calculate the corresponding spaceship coordinates x',ct' from Eqs. (14) and (15). Alternatively, we can find these coordinates graphically, by means of a carefully drawn Minkowski diagram, with calibrated axes and corresponding coordinate grids. Figure 2.15 shows such a Minkowski diagram for $V/c = 0.4$. The point $ct' = 1$ on the ct' axis is constructed by drawing the calibration curve $c^2t^2 - x^2 = 1$. The points $ct' = 2$, $ct' = 3$, etc. on the ct' axis are constructed by drawing the calibration curves $c^2t^2 - x^2 = 4$, $ct^2 - x^2 = 9$, etc. Of course, the points $ct' = 2$, $ct' = 3$, etc. can also be constructed by marking off equal intervals along the ct' axis. The points $x' = 1$, $x' = 2$, etc. on the x' axis can be constructed by drawing orthogonal calibration curves, $x^2 - ct^2 = 1$, $x^2 - c^2t^2 = 4$, etc. Alternatively, these points can be constructed by exploiting the symmetry between the x' and ct' axes.

For a graphical determination of the spacetime coordinates x,ct and x',ct' of any event P, we can use the coordinate grids shown in Fig. 2.15 or we can draw intercepts toward the appropriate axes. For the slanted x',ct' axes, the intercepts are slanted lines parallel to the axes.

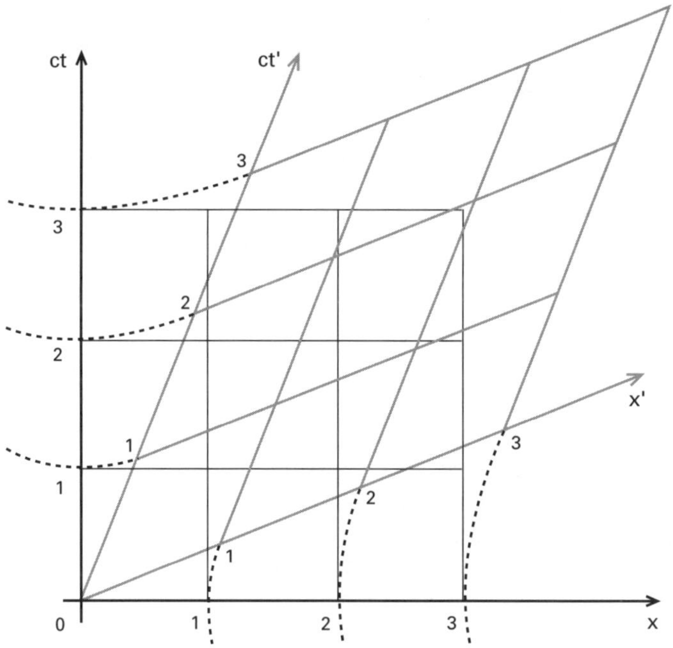

Fig. 2.15 Minkowski diagram with coordinate grids for a Lorentz transformation with $V = 0.4c$. The calibration of the ct' axis is fixed by the hyperbola $c^2t^2 - x^2 = 1$ connecting the point 1 on the ct axis with the point 1 on the ct' axis. The other hyperbolas, $c^2t^2 - x^2 = 4$, $c^2t^2 - x^2 = 9$, etc., connect other points on these axes, as indicated by the dashed curves. The coordinates x, ct and x', ct' of an event P can be determined graphically by means of the coordinate grids; alternatively, they can be determined by drawing intercepts toward the axes.

By analogy with the position vector of a point in ordinary space, we can define the spacetime position vector of an event P as the directed line segment OP in a spacetime diagram (see Fig. 2.16). The spacetime coordinates ct and x are then the components of the position vector. Thus, in the laboratory frame, the spacetime position vector has the components ct and x, whereas in the spaceship frame, it has the components ct' and x'. If we take the y and z coordinates into account, **the spacetime position vector** is a four-component vector, with components ct,x,y,z in the laboratory frame and components ct',x',y',z' in the spaceship frame. In the Minkowski diagram the spacetime position vector remains unchanged when we perform a Lorentz transformation, but its components change because the coordinate axes change.

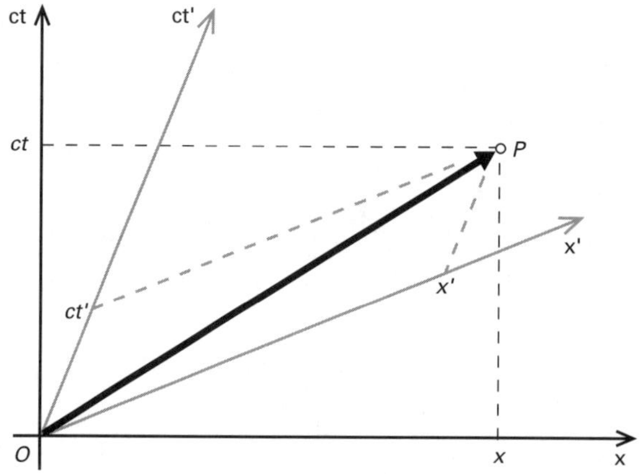

Fig. 2.16 An event P and its spacetime position vector OP. The components of this position vector are ct, x in the laboratory frame and ct', x' in the spaceship frame.

The Lorentz transformations treat space and time symmetrically. A change of reference frame intimately mixes space and time. What is a pure space displacement or a pure time displacement in one refence frame becomes a mixture of space and time displacements in another reference frame. In view of this intimate mixing of space and time, Minkowski declared that "henceforth space by itself, and time by itself, are doomed to fade away into mere shadows, and only a union of the two will preserve an independent reality" [4]. This union of space and time is the relativistic spacetime, with a geometry characterized by an invariant spacetime interval, which we will examine in the next section.

Note that if the speed V is low, we can neglect the denominators in Eqs. (14) and (15), so

$$x' \simeq x - Vt \tag{20}$$

$$t' \simeq t - Vx/c^2 \tag{21}$$

If the velocity V and the distance x are of the magnitudes typically encountered in laboratory experiments or observations on Earth, then the term Vx/c^2 can usually also be neglected, and then Eqs. (20) and

(21) reduce to the Galilean transformation equations,

$$x' = x - Vt \tag{22}$$

$$t' = t \tag{23}$$

Thus, as expected, the differences between the Lorentz transformation and the Galilean transformation are insignificant at low velocities.

But keep in mind that what is or is not significant depends on the precision of the measurements. For instance, if the "spaceship" reference frame is actually the reference frame of a supersonic aircraft flying at twice the speed of sound, $V \simeq 700$ m/s, and if x is the diameter of the Earth, $x \simeq 10^7$ m, then last term in Eq. (21) is $Vx/c^2 \simeq 700$ m/s $\times$ 10^7 m/(3 $\times$ 10^8 m/s)$^2 \simeq 10^{-7}$ s. This is a small time interval, too small to be detected by the onboard clock typically installed in the instrument panel of an aircraft. However, a time interval of 10^{-7} s is readily measurable with a portable atomic clock. This example illustrates that although the differences between relativistic and Newtonian physics become small when the speed is small compared with the speed of light, these differences never disappear entirely (except at zero speed), and whether they are detectable or not depends on how good our measuring instruments are.

■ **EXAMPLE 1.** An automobile collision occurs at exactly 10h 0m 0s Eastern Standard Time in Philadelphia, and another automobile collision occurs at exactly 10h 0m 0s in Boston, at a distance of 430 km from the first collision. The collisions are simultaneous in the reference frame of the Earth. What is the time difference between these collisions as measured in the reference frame of a spaceship traveling at $V = 0.9c$ in the direction from Philadelphia to Boston?

SOLUTION. Assume that the x axis is aligned with the two cities, with Philadelphia at $x_1 = 0$, and Boston at $x_2 = 4.3 \times 10^5$ m. In the reference frame of the Earth, the collisions occur simultaneously at $t_1 = t_2 = 10$ h. In the reference frame of the spaceship, they occur at

$$t_1' = \frac{t_1 - Vx_1/c^2}{\sqrt{1 - V^2/c^2}} = \frac{t_1}{\sqrt{1 - V^2/c^2}}$$

and at

$$t_2' = \frac{t_2 - Vx_2/c^2}{\sqrt{1 - V^2/c^2}} = \frac{t_1 - Vx_2/c^2}{\sqrt{1 - V^2/c^2}}$$

The time difference is

$$t_2' - t_1' = -\frac{Vx_2/c^2}{\sqrt{1 - V^2/c^2}} = -\frac{0.9 \times 4.3 \times 10^5 \text{ m/}(3 \times 10^8 \text{ m/s})}{\sqrt{1 - 0.9^2}} = -3.0 \times 10^{-3} \text{ s}$$

The negative sign indicates that the Philadelphia collision occurs later than the Boston collision, in agreement with the qualitative discussion in Section 2.2.

Although in this example the time difference is fairly small, a larger distance between the sites of the collisions would give a larger time difference. For instance, if one of the collisions happens on the Earth and the other on a planet orbiting a star several light-years away, the time difference generated by transforming to the reference frame of the spaceship would be several years!

It is important to understand that the time and space coordinates used in the Lorentz transformation equations are the time and space coordinates determined by measurements with the clocks and measuring rods of the reference frames. These are not necessarily the times and distances that one observer or another perceives with her eyes. For instance, an observer stationed near the site of the Phildelphia collision of Example 1 will perceive that collision first, and she will perceive the Boston collision later, when the light from the Boston collision arrives at her eye (she needs a good telescope to detect this light). The Lorentz transformations deal not with such subjective perceptions of different observers, but with the objective determinations of the time and space coordinates of events by means of reference frames.

2.4 THE SPACETIME INTERVAL

In the preceding section we found the equation for the calibration curve that connects the unit points on the old and the new time axes. In Eq. (9), we expressed the equation of this calibration curve in terms of the old coordinates x,ct. But we can also express the equation of this calibration curve in terms of the new coordinates x',ct'. In terms of these new coordinates, the equation has a similar form:

$$c^2t'^2 - x'^2 = 1 \tag{24}$$

This is so because instead of starting with the unit point on the ct axis and constructing the calibration curve by infinitesimal increments of the velocity in the old coordinates [as we did in the derivation leading to Eq. (9)], we can equally well start with the unit point on the ct' axis and construct the calibration curve by infinitesimal velocity increments in the new coordinates. The derivation then proceeds in the same way, and the result is an equation similar to $c^2t^2 - x^2 = 1$, but with primes, that is, $c^2t'^2 - x'^2 = 1$.

By comparing these two versions the calibration curve, Eqs. (9) and (24), we see that

$$c^2t^2 - x^2 = c^2t'^2 - x'^2 \tag{25}$$

This equality is of general validity, that is, whenever $c^2t^2 - x^2$ has some some arbitrary numerical value, $c^2t'^2 - x'^2$ has the same numerical value. For instance, if $c^2t^2 - x^2 = 5$, we can draw a calibration curve that starts at $ct = \sqrt{5}$. This curve intersects the ct' axis at $ct' = \sqrt{5}$, which tells us that, expressed in the new coordinates x',ct', the equation of this calibration curve is $c^2t'^2 - x'^2 = 5$. Thus $c^2t^2 - x^2$ has the same numerical value as $c^2t'^2 - x'^2$. The equality (25) can also be verified by direct substitution of ct' and x' according to the Lorentz transformation equations (see Problem 18).

The invariance of $c^2t^2 - x^2$ under Lorentz transformations is analogous to the invariance of $x^2 + y^2$ under ordinary rotations in the x-y plane. According to ordinary Euclidean geometry, $x^2 + y^2$ is the square of the distance between the origin and the point x,y, and ordinary rotations of the coordinates leave this distance invariant. This analogy suggests that $c^2t^2 - x^2$ is some kind of relativistic "distance" between the origin and the point x,ct. This "distance" is called the spacetime interval. (For more on the analogy between Lorentz transformations and rotations, see Problems 14 and 15.)

More generally, consider two spacetime points t_1,x_1 and t_2,x_2 separated by a spacetime displacement $\Delta t = t_2 - t_1$, $\Delta x = x_2 - x_1$. The **spacetime interval** Δs between these spacetime points is defined as

$$\Delta s = \sqrt{(c\Delta t)^2 - (\Delta x)^2} \tag{26}$$

If $(c\Delta t)^2 > (\Delta x)^2$, the spacetime interval is real, but if $(c\Delta t)^2 < (\Delta x)^2$, it is imaginary. Since imaginary quantities are a nuisance, it is convenient to deal with the square of the spacetime interval,

$$(\Delta s)^2 = (c\Delta t)^2 - (\Delta x)^2 \tag{27}$$

This is positive if $(c\Delta t)^2 > (\Delta x)^2$ and negative if $(c\Delta t)^2 < (\Delta x)^2$.

The spacetime interval between any two spacetime points is invariant under Lorentz transformations, that is, if $(\Delta s)^2$ has some numerical value in one inertial reference frame, then it has the same numerical value in any other inertial reference frame:

$$(c\Delta t)^2 - (\Delta x)^2 = (c\Delta t')^2 - (\Delta x')^2 \tag{28}$$

In Eq. (25) we established this invariance for the special case of a displacement from the origin to the point t,x, for which $t_1 = 0$, $x_1 = 0$, $t_2 = t$, $x_2 = x$, and $\Delta t = t$, $\Delta x = x$, so $(c\Delta t)^2 - (\Delta x)^2 = (ct)^2 - x^2$. This special case is sufficient to ensure the validity of the general equation (28), since we can always place the origin at t_1,x_1, and thereby reduce the displacement to the special case.

In the above discussion of the spacetime interval, we have omitted the y and z coordinates. But the equation for the invariance of the spacetime interval remains valid even when we include these coordinates. The square of the spacetime interval is then

$$(\Delta s)^2 = (c\Delta t)^2 - (\Delta x)^2 - (\Delta y)^2 - (\Delta z)^2 \tag{29}$$

The distances transverse to the velocity are unaffected by the Lorentz transformation [see Eqs. (18) and (19)], so including the y and z coordinates does not affect the invariance of the spacetime interval:

$$(c\Delta t)^2 - (\Delta x)^2 - (\Delta y)^2 - (\Delta z)^2 = (c\Delta t')^2 - (\Delta x')^2 - (\Delta y')^2 - (\Delta z')^2 \tag{30}$$

In the preceding section we defined a spacetime position vector with components ct, x, y, z. Likewise, we can define a spacetime displacement vector with components $c\Delta t$, Δx, Δy, Δz. The spacetime interval Δs can be regarded as the "length" of the spacetime displacement vector. The invariance of this "length" under a Lorentz transformation is then analogous to the invariance of the length of a vector in ordinary space under a transformation of the x,y,z coordinate axes, such as a rotation of the x,y,z coordinate axes.

The spacetime interval plays an important role in relativity, because it characterizes the geometry of relativistic spacetime. The crucial distinction between the geometry of relativistic spacetime and the geometry of Newtonian spacetime is that in the latter, both the length interval Δl (at a given time) and the time interval Δt are invariant; whereas in the former, only the spacetime interval $\Delta s = \sqrt{(c\Delta t)^2 - (\Delta l)^2}$ is invariant (see Table 2.1).

TABLE 2.1 COMPARISON OF INVARIANTS

	Newtonian Spacetime	**Relativistic Spacetime**
Coordinates of a Spacetime Point	t,x,y,z	t,x,y,z
Invariants	$(\Delta l)^2 = (\Delta x)^2 + (\Delta y)^2 + (\Delta z)^2$ Δt	$(\Delta s)^2 = c^2(\Delta t)^2 - (\Delta x)^2 - (\Delta y)^2 - (\Delta z)^2$

The geometry of Newtonian spacetime consists of two disjoint Euclidean geometries for space and for time, that is, a three-dimensional geometry for space and a one-dimensional geometry for time. The geometry of relativistic spacetime consists of a single four-dimensional pseudo-Euclidean geometry. The term *pseudo-Euclidean* refers to the presence of negative signs in Eq. (29), which means that the square of the spacetime interval is not positive definite.

In a purely formal way, it is possible to define an imaginary time coordinate $\chi = ict$, in terms of which Eq. (29) for the spacetime interval can be expressed as

$$-(\Delta s)^2 = (\Delta \chi)^2 + (\Delta x)^2 + (\Delta y)^2 + (\Delta z)^2 \tag{31}$$

Here the right side is a sum of squares, and the geometry of spacetime looks Euclidean. In view of this, it is sometimes said that time is an imaginary dimension of four-dimensional space. However, the imaginary coordinate χ in Eq. (31) hides a minus sign, and the real geometry of spacetime is still pseudo-Euclidean. Although the imaginary coordinate χ is sometimes helpful in the derivation of mathematical results in relativity (see, for example, Problem 14), we will not use it in this book.

The square of the spacetime interval (29) is positive if $(c\Delta t)^2 > (\Delta x)^2 + (\Delta y)^2 + (\Delta z)^2$ and negative if $(c\Delta t)^2 < (\Delta x)^2 + (\Delta y)^2 + (\Delta z)^2$. Intervals with a positive square are called **timelike** (because the time part predominates). Intervals with a negative square are called **spacelike** (because the space part predominates). And intervals that are zero are called **lightlike** (because they correspond to the worldlines of light signals). Figure 2.17 shows examples of displacements with timelike, spacelike, and lightlike spacetime intervals. Note that the angles between these displacements and the t axis are less than 45°, more than 45°, and equal to 45°, respectively.

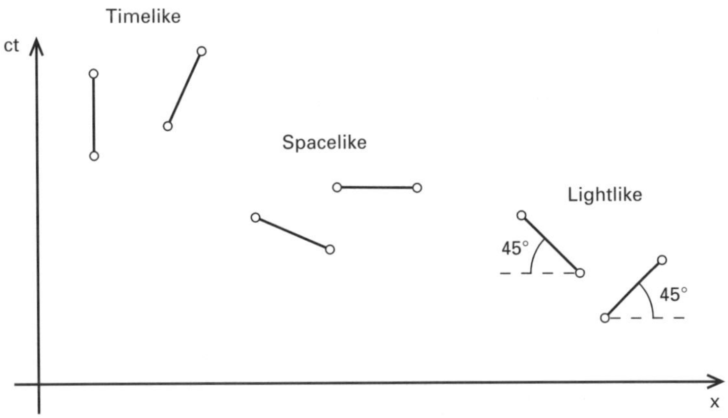

Fig. 2.17 Examples of displacements with timelike, spacelike, and lightlike spacetime intervals.

The set of all lightlike lines starting at the origin (that is, worldlines of light signals emitted at the origin) forms a three-dimensional "surface" in the four-dimensional spacetime. This surface has the equation $c^2t^2 - x^2 - y^2 - z^2 = 0$ and is called the **light cone**. We cannot visualize four-dimensional spacetime and draw the three-dimensional surface of the light cone. The best we can do is to omit one space coordinate, say the z coordinate. The light cone then reduces to the two-dimensional surface $c^2t^2 - x^2 - y^2 = 0$, which is an ordinary cone in three-dimensional space. Figure 2.18 shows this cone.

The upper ($t > 0$) and the lower ($t < 0$) parts of the cone are called, respectively, the **future** and the **past** light cones, or the forward and the backward light cones. Particles that leave the origin must have worldlines inside the future light cone; particles that arrive at the origin must have worldlines inside the past light cone.

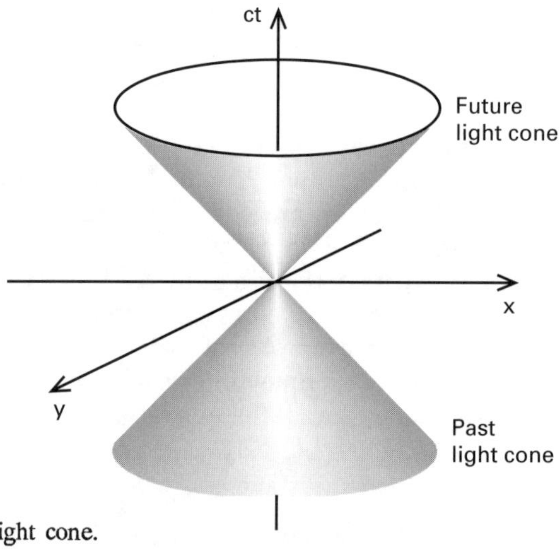

Fig. 2.18 The light cone.

EXAMPLE 2. At 7h 0m 0s Eastern Standard Time a meteoroid strikes the surface of the Moon, and at 7h 0m 1s another meteoroid strikes the surface of the Earth. In the reference frame of the Earth, the distance between the impact sites is 3.8×10^8 m. Is the spacetime interval spacelike or timelike? Find a reference frame in which the two events are simultaneous.

SOLUTION. Assume that the site on the Moon is at $x = 0$ and the site on the Earth at $x = 3.8 \times 10^8$ m. The spacetime interval between the events is

$$(\Delta s)^2 = c^2(\Delta t)^2 - (\Delta x)^2 = (3.0 \times 10^8 \text{ m/s})^2(1 \text{ s})^2 - (3.8 \times 10^8 \text{ m})^2$$
$$= -5.4 \times 10^{16} \text{ m}^2$$

Since this is negative, the spacetime interval is spacelike (see Fig. 2.19).

To make $\Delta t' = 0$ in the reference frame of a spaceship, we want

$$\Delta t' = \frac{\Delta t - V\Delta x/c^2}{\sqrt{1 - V^2/c^2}} = 0$$

which requires

$$V = \frac{\Delta t}{\Delta x} c^2 = \frac{1 \text{ s}}{3.8 \times 10^8 \text{ m}} \times (3.0 \times 10^8 \text{ m/s})^2$$
$$= 2.4 \times 10^8 \text{ m/s} = 0.79c$$

Thus, in the reference frame of a spaceship traveling from the Moon toward the Earth at 0.79*c*, the events are simultaneous.

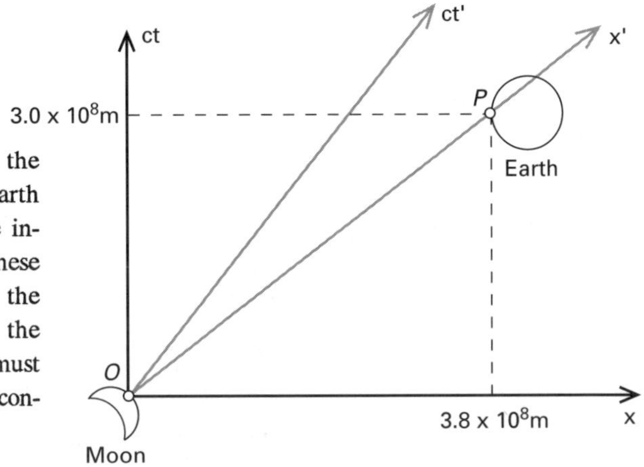

Fig. 2.19 Events on the Moon and on the Earth separated by a spacelike interval. To make these events simultaneous in the reference frame of the spaceship, the x' axis must coincide with the line connecting these events.

2.5 THE COMBINATION OF VELOCITIES

The Galilean addition law for velocities is a direct consequence of the absolute character of length and of time in Newtonian physics. Likewise, the relativistic combination law for velocities is a direct consequence of the relativistic character of length and time inherent in the Lorentz transformation equations.

Suppose that x,ct and x',ct' are the spacetime coordinates of a moving particle. The combination law for velocities follows immediately from the Lorentz transformation equations for differentials, obtained by considering small differences of the space and time coordinates x,t and x',t',

$$dx' = \gamma\left[dx - \frac{V}{c}\,\mathrm{cdt}\right] \tag{32}$$

$$dt' = \gamma\left[dt - \frac{V}{c^2}\,dx\right] \tag{33}$$

The ratio of these differentials gives

$$\frac{dx'}{dt'} = \frac{dx - Vdt}{dt - Vdx/c^2} = \frac{dx/dt - V}{1 - (V/c^2)dx/dt} \tag{34}$$

Here $dx/dt = v_x$ is the velocity of the particle with respect to the laboratory reference frame, and $dx'/dt' = v'_x$ is the velocity with respect to the spaceship frame. We can then rewrite Eq. (34) as

$$v'_x = \frac{v_x - V}{1 - Vv_x/c^2} \tag{35}$$

This relativistic combination law for velocities differs from the Galilean addition law by the denominator $(1 - Vv_x/c^2)$. This denominator can be ignored if the velocities v_x and V are low, so $Vv_x \ll c^2$. The relativistic combination law is then nearly identical to the Galilean addition law. But if the velocities are large, then these laws differ drastically. For instance, if v_x is the velocity of a light signal propagating along the x axis, then $v_x = c$ and Eq. (35) gives

$$v'_x = \frac{c - V}{1 - V/c} = c \tag{36}$$

Thus, the velocity of the light signal has exactly the same magnitude in the spaceship frame as in the laboratory frame. Because the construction of the Lorentz transformation equations incorporates the invariance of the speed of light, it was to be expected that the relativistic combination law for velocities should yield the speed of light whenever the speed of light is combined with any other speed, and Eq. (36) confirms this.

The inverse combination that expresses the velocity v_x in terms of v'_x is

$$v_x = \frac{v'_x + V}{1 + Vv'_x/c^2} \tag{37}$$

This result can be obtained either by solving Eq. (35) for v_x or by exchanging primed and unprimed quantities in Eq. (35) and changing V to $-$ V.

Equation (35) applies only to the x component of the velocity, that is, the component parallel to the relative motion of the reference frames. The y and the z components of the velocity have combination laws of their own, as follows:

$$v_y' = \frac{\sqrt{1 - V^2/c^2}}{1 - Vv_x/c^2} v_y \tag{38}$$

$$v_z' = \frac{\sqrt{1 - V^2/c^2}}{1 - Vv_x/c^2} v_z \tag{39}$$

These combination laws can be obtained by taking the ratio of the differentials dy' or dz' and dt'.

■ **EXAMPLE 3.** A high-energy neutron of speed 0.40c produced in an accelerator experiment decays by emitting a beta ray, or a fast electron. The speed of this electron is 0.92c in the rest frame of the neutron. What is the speed of the electron in the laboratory frame if it is emitted in the forward direction? in the backward direction?

SOLUTION. For forward emission ($v'_x = 0.92c$), the relativistic velocity combination formula (37) gives

$$v_x = \frac{v'_x + V}{1 + Vv'_x/c^2} = \frac{0.92c + 0.40c}{1 + 0.40 \times 0.92} = 0.96c$$

and for backward emission ($v'_x = -0.92c$), it gives

$$v_x = \frac{v'_x + V}{1 + Vv'_x/c^2} = \frac{-0.92c + 0.40c}{1 - 0.40 \times 0.92} = -0.82c$$

Experimentally, the velocity combination law has been tested with moving sources of light. According to Eq. (36), the light emitted by a moving source has the same speed c as the light from a stationary source. This independence of the speed of light from the speed of the source is, of course, also true in the ether theory. Hence an experimental test of this independence cannot discriminate between relativity and ether theory, but it can discriminate between relativity and emission theories in which a source with a speed V produces light with a speed $c \pm V$ (+ if the source is approaching, - if the source is receding). As we noted in Chapter 1, a dependence of the speed of light on the speed of the source would lead to the observation of weird effects in binary systems of stars. In such systems, the two stars that are the sources of the light have different velocities in their orbits around each other. According to emission theories, this would lead to different speeds for their light and to different travel times for this light to reach the Earth. Consequently, an astronomer on the Earth would see these stars as they were at different times in the past, which would result in an apparent irregular behavior of their apparent orbital motion. In an analysis of observational data on the orbits of binary stars, W. de Sitter found no such irregularities, and he concluded that the alteration of the speed of light is at most $c \pm 0.002V$, that is, about 500 times smaller than expected according to emission theory [5].

Further confirmation of the relativistic velocity combination law emerged from an analysis of observational data on X-ray pulses emitted by pulsating binary X-ray sources, such as Hercules X-1 [6]. This source consists of a neutron star, which is orbiting around a normal star. The component of the velocity of the neutron star along our line of sight changes by about 300 km/s as this star moves around its orbit. If this change of velocity were to affect the speed of the X rays emitted by the star, then pulses from different portions of the orbit would travel toward the Earth at different speeds, and some pulses emitted at different times would arrive at the Earth at the same time. Thus, an observer on the Earth would see several sets of overlapping pulses, or "ghosts." From the absence of such overlapping pulses in the observational data, it is possible to infer an upper limit on the variation of the speeds of X rays caused by the orbital motion: the speeds of X rays emitted at different times vary by no more than 6×10^{-7} km/s, compared with a variation of 300 km/s expected according to emission theory.

The relativistic combination law for velocities has also been tested in a laboratory experiment using a beam of very fast neutral π mesons produced by the CERN accelerator [7]. These particles decay spontaneously into two gamma rays (two high-energy photons); thus, such a beam of π mesons can be regarded as a high-speed light source. In the experiment, the speed of the π mesons relative to the laboratory was $V = 0.99975c$. Emission theory would then predict a laboratory velocity of $v_x = 1.99975c$ for gamma rays emitted in the forward direction and a velocity of $v_x = 0.00025c$ for gamma rays emitted in the backward direction. But the experiment confirmed the relativistic combination law--the laboratory velocity of the gamma rays was the same in both directions.

2.6 *MOTION WITH CONSTANT ACCELERATION*

A characteristic feature of the Galilean transformation equations is the invariance of the acceleration: if a particle has a given acceleration in one reference frame, then it has the same acceleration in any other reference frame. In contrast, the Lorentz transformations do not preserve the acceleration, that is, the acceleration is different in different reference frames.

The relativistic transformation equations for the acceleration can be obtained by examining the differentials of the transformation equations for the velocity. For the x component this yields

$$dv'_x = \frac{dv_x}{1 - Vv_x/c^2} + \frac{v_x - V}{(1 - Vv_x/c^2)^2}\frac{V}{c^2}\,dv_x = \frac{1 - V^2/c^2}{(1 - Vv_x/c^2)^2} \tag{40}$$

With $dt' = dt(1 - Vv_x/c^2)/\sqrt{1 - V^2/c^2}$, and $a_x = dv_x/dt$, $a'_x = dv'_x/dt'$, we find

$$a'_x = a_x\frac{(1 - V^2/c^2)^{3/2}}{(1 - Vv_x/c^2)^3} \tag{41}$$

Similarly,

$$a'_y = a_y \frac{1 - V^2/c^2}{(1 - Vv_x/c^2)^2} + a_x(1 - V^2/c^2)\frac{Vv_y/c^2}{(1 - Vv_x/c^2)^3} \tag{42}$$

$$a'_z = a_z \frac{1 - V^2/c^2}{(1 - Vv_x/c^2)^2} + a_x(1 - V^2/c^2)\frac{Vv_z/c^2}{(1 - Vv_x/c^2)^3} \tag{43}$$

If a particle has constant acceleration a_x in the laboratory frame, then its velocity will be $v_x = a_x t$, and its position will be $x = \frac{1}{2} a_x t^2$, as in the nonrelativistic case. However, because the particle cannot exceed the speed of light, the acceleration must cease when $v_x = c$. (In Chapter 3 we will see how the particle becomes more and more difficult to accelerate as it approaches the speed of light, and we will see that the acceleration indeed ceases if the force acting on the particle remains finite.)

A more interesting case is a particle with constant acceleration in its own rest frame, more precisely, a particle that has a constant acceleration in its instantaneous inertial rest frame. The instantaneous inertial rest frame has the same velocity as the particle at one instant, but it has no acceleration; thus, for each instant, there is a different instantaneous inertial frame. While the particle accelerates, it "jumps" from one instantaneous inertial rest frame to another. The velocity of the instantaneous inertial rest frame coincides with the velocity of the particle, so $V = v_x$, and according to Eq. (41), the acceleration of the particle in this instantaneous rest frame becomes

$$a'_x = \frac{dv_x}{dt}\frac{(1 - v_x{}^2/c^2)^{3/2}}{(1 - v_x{}^2/c^2)^3} = \frac{dv_x}{dt}\frac{1}{(1 - v_x{}^2/c^2)^{3/2}} \tag{44}$$

With a constant value of a'_x, this equation can be easily integrated:

$$a'_x t = \int \frac{dv_x}{(1 - v_x{}^2/c^2)^{3/2}} = \frac{v_x}{\sqrt{1 - v_x{}^2/c^2}} + \text{constant} \tag{45}$$

We will adopt the initial condition $v_x = 0$ at $t = 0$, so the constant equals zero. If we then solve Eq. (45) for v_x, we find

$$v_x = \frac{a'_x t}{\sqrt{1 + a'_x{}^2 t^2/c^2}} \tag{46}$$

This, again, can be easily integrated

$$x = \int \frac{a'_x t}{\sqrt{1 + a'^2_x t^2/c^2}} dt = \frac{c^2}{a'_x} \sqrt{1 + a'^2_x t^2/c^2} + \text{constant} \qquad (47)$$

With the initial condition $x = 0$ at $t = 0$, the constant equals $- c^2/a'_x$, and

$$\boxed{x = \frac{c^2}{a'_x} \left(\sqrt{1 + a'^2_x t^2/c^2} - 1\right)} \qquad (48)$$

For short times ($t \ll c/a'_x$), this coincides approximately with the familiar nonrelativistic formula $x = \frac{1}{2} a'_x t^2$, but for longer times, and higher speeds, it differs drastically (see Fig. 2.20). Note that for $t \to \infty$, the velocity v_x asymptotically approaches the speed of light but never reaches it.

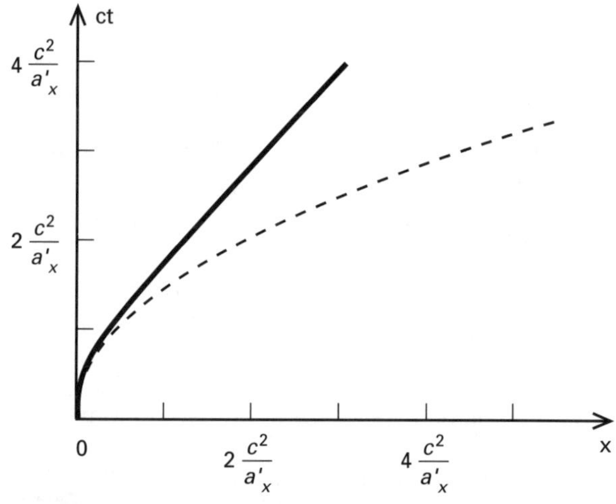

Fig. 2.20 Worldline of an accelerated particle according to the relativistic formula (48). The slope of this worldline approaches 1 asymptotically, corresponding to the limiting speed c. For comparison, the dashed line gives the worldline according to the nonrelativistic formula $x = \frac{1}{2} a'_x t^2$.

The motion of a charged particle placed in a uniform electric field is a practical example of motion with constant acceleration in the instantaneous inertial rest frame. For a particle of charge e and mass m in an electric field E, the acceleration in the rest frame has the constant value $a'_x = eE/m$. Thus, the particle moves according to Eq. (48) [however, if we take into account the radiation emitted by such an accelerated particle, the motion differs somewhat from Eq. (48)].

The motion of a rocket with constant thrust is *not* motion with constant acceleration. As the rocket expends its fuel, its mass decreases, and the acceleration therefore increases. However, it is possible to program the rocket engines in such a way that the thrust decreases in the same way as the mass, and it is then possible to achieve constant acceleration in the instantaneous inertial rest frame. Such a constant acceleration in the rest frame of a rocket would be of interest to astronauts of the future. The astronauts could maintain a constant acceleration a'_x equal to the normal acceleration of gravity (9.8 m/s²) and could thereby simulate the normal gravity on the surface of the Earth.

■ **EXAMPLE 4.** Suppose that a rocket accelerates with a constant acceleration $a'_x = 9.8\ \mathrm{m/s^2}$ in its own rest frame. If this acceleration lasts for 1 year of Earth time, what is the final speed? What would be the final speed if Newtonian physics were valid?

SOLUTION. According to Eq. (46),

$$v_x = \frac{a'_x t}{\sqrt{1 + a'^2_x t^2/c^2}} = \frac{9.8\ \mathrm{m/s^2} \times 3.2 \times 10^7\ \mathrm{s}}{\sqrt{1 + (9.8\ \mathrm{m/s^2} \times 3.2 \times 10^7\ \mathrm{s}/3.0 \times 10^8\ \mathrm{m/s})^2}}$$

$$= 2.2 \times 10^8\ \mathrm{m/s}$$

According to Newtonian physics, the final speed would be

$$v_x = a'_x t = 9.8\ \mathrm{m/s^2} \times 3.2 \times 10^7\ \mathrm{s} = 3.1 \times 10^8\ \mathrm{m/s}$$

By an odd coincidence, the speed calculated according to Newtonian physics is almost exactly equal to the speed of light.

REFERENCES

1. A. Einstein, in *Albert Einstein: Philosopher-Scientist*, edited by P. A. Schilpp (Harper, New York, 1949), p. 53.
2. A. R. Lee and T. M. Kalotas, Am. J. Phys. **43**, 434 (1975).
3. J.-M. Lévy-Leblond, Am. J. Phys. **44**, 271 (1976).
4. H. A. Lorentz, A. Einstein, H. Minkowski, and H. Weyl, *The Principle of Relativity* (Methuen, New York, 1923), p. 75.
5. W. de Sitter, Proc. Amsterdam Acad. **16**, 395 (1913).
6. K. Brecher, Phys. Rev. Lett. **39**, 1051 (1977).
7. T. Alväger, F. J. M. Farley, J. Kjellman, and I. Wallin, Phys. Lett. **12**, 260 (1964).

PROBLEMS

1. Suppose that we send a spaceship on a trip to Alpha Centauri, 4 light-years away. The spaceship travels at a constant speed of $0.5c$, it stops at Alpha Centauri for 1 year, and it then returns to the Earth at constant speed $0.5c$.
(a) Draw a spacetime diagram with axes x, ct and plot the worldline of the spaceship.
(b) Suppose that 1 year after the departure of the spaceship, we send a radio message from the Earth toward the ship. Plot the worldline of this radio signal. At

what time does it reach the spaceship?

2. A spaceship moves at a speed of $0.3c$ along the positive x axis relative to the Earth.
(a) For this speed, draw the Minkowski diagram with the coordinate grids x, ct and x', ct' as in Fig. 2.15. Be sure to draw the correct angle for the x', ct' axes and to draw the correct calibration for these axes.
(b) On your diagram, mark the spacetime point $x = 2$, $ct = 4$. Determine the values of x' and ct' for this point graphically, by means of your coordinate grid.
(c) On your diagram, plot the worldline of an electron moving with velocity $v_x = 0.5c$. What is the slope of this worldline relative to the x, ct axes? Graphically determine the slope of the worldline relative to the x', ct' axes and thereby find v'_x.
(*d*) Compare the values of your graphical determinations in (b) and (c) with the numerical computations of these values from Eqs. (12), (13), and (35).

3. Consider a spaceship that moves at a speed of $0.4c$ along the *negative* x axis relative to the Earth. Draw the Minkowski diagram for this case. How does it differ from the Minkowski diagram for a speed along the positive x axis?

4. Astronomers on the Earth (regarded as an inertial reference frame) see two novas flare up simultaneously. One of the novas is at a distance of 1.0×10^3 light-years in the constellation Draco; the other nova is at an equal distance in the constellation Tucana in a direction (as see from the Earth) exactly opposite to that of the first nova. According to astronomers aboard an aircraft traveling at 750 km/h along the line from Draco to Tucana, the novas are not simultaneous. According to these astronomers, which nova happened first? By how many hours?

5. A spaceship has a speed of $0.80c$ relative to the Earth. In its own reference frame, the length of this spaceship is 300 m.
(a) Consider a pulse of light emitted from the tail of this spaceship. In the reference frame of the spaceship, how long does this pulse take to reach the nose?
(b) In the reference frame of the Earth, how long does this take? Calculate this time directly from the motions of the spaceship and the light pulse; then recalculate it by applying the Lorentz transformations to the result obtained in (a).

6. At 14h 0m 0s lightning strikes in New York City; and at 14h 0m 0.003s lighting strikes in Miami, at a distance of 1760 km from the first stroke. Find the velocity, relative to the Earth, of a spaceship in which the first stroke of lightning occurs after the second.

7. (a) Suppose that we install an array of mirrors along the equator of the Earth, so as to compel a light ray to travel along the equator all the way around the Earth. From one point on the equator, we send a light signal westward around the Earth and another light signal eastward around the Earth. Taking into account the rotation of the Earth, which of these signals returns to the starting point first? By how many seconds?
(b) The result of part (a) is in conflict with a universal, constant value of the speed

of light in the reference frame of the ground, and it tells us that on the rotating Earth we cannot synchronize clocks by light signals, according to the standard procedure described in Section 2.2. Does this contradict the Principle of Relativity?

8. Suppose that we place a sequence of n clocks along the equator of the Earth, with small distances Δx between one clock and the next. Since the Earth rotates, these clocks are not in an inertial reference frame, and they cannot be synchronized by the standard procedure with light signals (compare the preceding problem). However, two adjacent clocks have approximately the same rotational velocity $V = \omega R$ (in magnitude and direction), and they can be regarded as having an instantaneous uniform eastward translational motion relative to the inertial reference frame centered on the Earth (and not rotating with it). This suggests that to synchronize the clocks, we ought to adjust them in such a way that there is a time shift $-V\Delta x/\sqrt{1 - V^2/c^2}/c^2 \simeq -V\Delta x/c^2 = -\omega R\Delta x/c^2$ between adjacent clocks as observed from the inertial reference frame.

(a) If we proceed eastward around the Earth and successively adjust adjacent clocks according to this prescription, what is the net time shift between the second clock and the first? between the third clock and the first? between the nth clock and the first? What is the numerical value of this last time shift?

(b) What is the time shift between the first clock and the nth if instead of proceeding by $n - 1$ steps eastward, we proceed by 1 step westward?

(The conflict between the results obtained in (a) and (b) implies that it is not possible to attain synchronization in the rotating reference frame by pretending adjacent clocks have an instantaneous uniform translational motion. In practice, physicists avoid this synchronization conflict in the rotating reference frame of the Earth by giving the clocks on the rotating Earth the *same* synchronization as the clocks of the nonrotating inertial reference frame centered on the Earth; that is, they simply ignore the time shift associated with the rotational motion.)

9. In the laboratory frame, a rigid wheel of radius R spins with angular velocity ω about a horizontal axis (x axis). Consider this wheel from the point of view of a spaceship moving along the positive x axis at speed V.

(a) What is the transverse velocity of a point on the rim of the wheel in the spaceship frame?

(b) What is the angular velocity of rotation of the wheel in the spaceship frame?

10. Two wheels are welded to the ends of a rigid straight shaft of length L aligned with the x axis (see Fig. 2.21). According to the manufacturer's specifications, in the rest frame of the shaft, corresponding spokes of the wheels are parallel, as indicated by the parallel dashed lines in the figure. Suppose that in the laboratory frame, the shaft and wheels have no translational motion, but they rotate rigidly with angular velocity ω. In a spaceship frame moving toward the left at speed V (in which the wheels move toward the right at the same speed V), what is the angular displacement between the corresponding spokes of the two wheels measured at one instant of t' time?

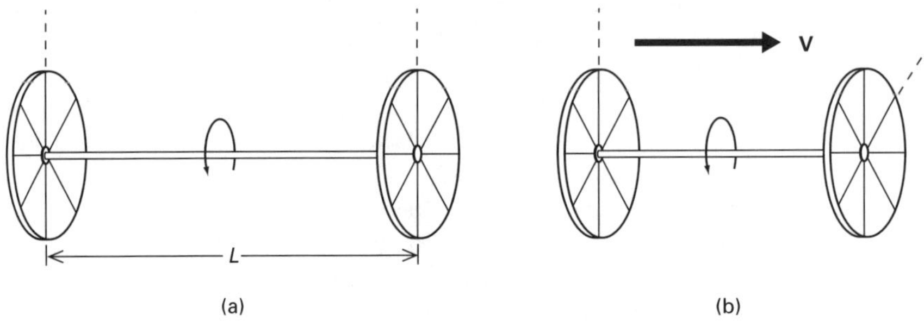

Fig. 2.21 Two wheels welded to a rigid shaft. (a) In the laboratory frame the shaft is at rest, and the spokes of the wheels are aligned. (b) In the spaceship frame the shaft moves toward the right at speed V, and at one instant of t' time, the spokes are misaligned.

11. A straight rigid rod of length $2L$ lies in the x-y plane and rotates about the z axis with uniform angular velocity ω. In the x,y,z frame, the rod is perfectly straight at each instant of time, and the points on the rod can be described parametrically by

$$x = R \cos \omega t$$

$$y = R \sin \omega t$$

where R is a parameter that ranges from $-L$ to L. Consider this rod from the point of view of the x',y',z' frame moving along the x axis with velocity V. Where are the points on the rod at one instant of t' time? Is the rod still straight? Roughly sketch the shape of the rod at one instant of t' time, say, $t' = 0$; assume that $V = 0.8c$ and $\omega L = 0.8c$.

12. Show that the combination of two successive Lorentz transformations with velocities V_1 and V_2 along the x axis is equivalent to a single Lorentz transformation with a velocity

$$V = \frac{V_1 + V_2}{1 + V_1 V_2 / c^2}$$

13. As mentioned in Section 2.3, the inverse Lorentz transformation equations can be obtained in two ways: by solving the equations

$$x' = \gamma(x - Vt)$$

$$ct' = \gamma(ct - Vx/c)$$

for x,t or by simply exchanging primed and unprimed coordinates and replacing V by $-V$. Apply both of these methods to these equations, treating γ as an unknown constant. By comparing the resulting equations for x',t', deduce the value of γ.

14. (a) Show that the Lorentz transformation can be written as

$$x' = x \cosh\theta - ct \sinh\theta$$

$$ct' = -x \sinh\theta + ct \cosh\theta$$

where θ is a "velocity parameter" defined by $V/c = \tanh\theta$.
(b) Show that in terms of the imaginary time coordinate $\chi = ict$, this can also be written as

$$x' = x \cos(i\theta) + \chi \sin(i\theta)$$

$$\chi' = -x \sin(i\theta) + \chi \cos(i\theta)$$

These equations are formally identical to the equations for the transformation of the coordinates under a rotation in the x-χ plane; but the angle of the rotation is $i\theta$, which is imaginary.
(c) Show that the combination of two successive Lorentz transformations with velocity parameters θ_1 and θ_2 is equivalent to a single Lorentz transformation with velocity parameter $\theta = \theta_1 + \theta_2$. [*Hint*: Use the identity for the hyperbolic tangent of a sum of two angles, $\tanh(\theta_1 + \theta_2) = (\tanh\theta_1 + \tanh\theta_2)/(1 + \tanh\theta_1 \tanh\theta_2)$].

15. Instead of describing events in spacetime by the spacetime coordinates x, t, we can use other spacetime coordinates. One such set of alternative coordinates consists of the **Rindler coordinates** X, T. In the "wedge" $|x| > ct$ (see Fig. 2.22), these coordinates are defined by

$$X = \sqrt{x^2 - c^2t^2}$$

$$T = \tanh^{-1} ct/x$$

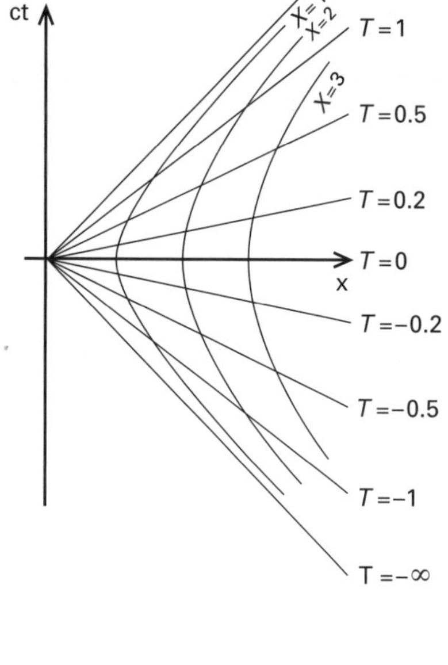

Fig. 2.22 The grid of Rindler coordinates X, T in the wedge $|x| > ct$.

In the other wedges of the x-t plane, the definitions are slightly different. The curves X = constant are hyperbolas, and the curves T = constant are straight lines through the origin. Show that in terms of X,T the Lorentz transformation takes the very simple form

$$X' = X$$

$$T' = T - \theta$$

where $\theta = \tanh^{-1} V/c$. [*Hint*: Begin with Eqs. (12) and (13), and use $(\tanh\alpha - \tanh\beta)/(1 - \tanh\alpha \tanh\beta) = \tanh(\alpha - \beta)$.] Note that if we think of X as a radial coordinate and of T as an angular coordinate, then the transformation equations for X and T are analogous to a rotation: the "radial" coordinate X remains unchanged, and the "angular" coordinate T changes by a constant amount θ.

16. Figure 2.23 is a **Loedel diagram**, a graphical representation of the Lorentz transformation which is an alternative to the Minkowski diagram. In the Loedel diagram, both pairs of axes x, ct and x', ct' are slanted; the x axis is perpendicular to the ct' axis, and the x' axis is perpendicular to the ct axis. The slant angle α is choosen in such a way that the coordinates that are read off the diagram are related by the Lorentz transformation. Show that if the angle α in Fig. 2.23 is given by

$$\sin\alpha = V/c$$

$$\cos\alpha = \sqrt{1 - V^2/c^2}$$

then the coordinates determined from the diagram by means of the dashed lines shown in Fig. 2.23 agree with the Lorentz transformations, Eqs. (12) and (13). Compared with the Minkowski diagram, the Loedel diagram has the advantage that the calibrations of the ct and ct' axes are the same (the unit points do not need to be adjusted by any calibration curve). However, it has the serious disadvantage that any alteration of the velocity V requires a rearrangement of the ct axis and a replotting of all events and all worldlines. Because events and worldlines represent *real* things, it is desirable that they should remain fixed when we change our coordinates, as they do in the Minkowski diagram.

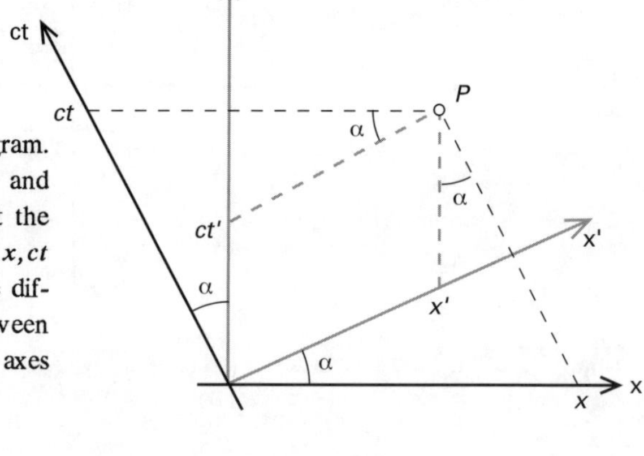

Fig. 2.23 Loedel diagram. Both the sets of axes x, ct and x', ct' are slanted. Note that the slant angles between the x, ct axes and the x', ct' axes are different, but the angles between the x, x' axes and the ct, ct' axes are equal.

17. In the reference frame of the Earth, a meter rod is placed at rest in the x-y plane. The meter rod makes an angle ϕ with the x axis (see Fig. 2.24). What angle ϕ' does this meter rod make with the x' axis of a spaceship traveling at speed V along the x axis of the Earth frame? Evaluate numerically for $\phi = 30^0$ and $V = 0.8c$.

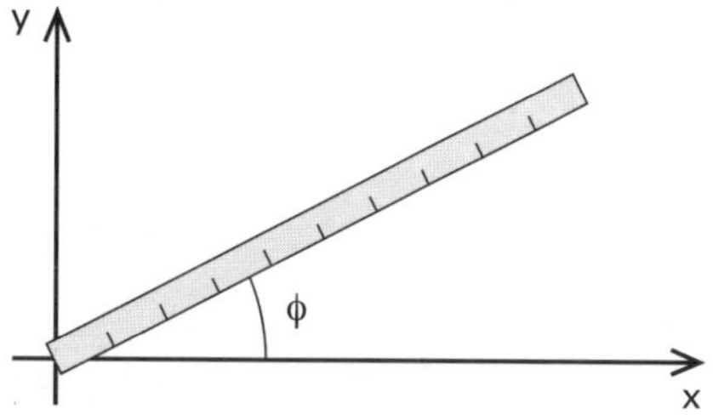

Fig. 2.24 A meter rod in the x-y plane, inclined at an angle ϕ.

18. By direct substitution of x' and ct' according to the Lorentz transformation equations (12) and (13), verify that $ct'^2 - x'^2 = ct^2 - x^2$.

19. Consider two spacetime position vectors, with components ct_1, x_1, y_1, z_1 and ct_2, x_2, y_2, z_2, respectively. Show that the "scalar" product of these vectors, defined as $c^2 t_1 t_2 - x_1 x_2 - y_1 y_2 - z_1 z_2$, is invariant under Lorentz transformations.

20. In the laboratory frame, two events are separated by a time interval of 8.0 s and a spatial distance of 2.0×10^9 m. Does there exist a spaceship frame in which the two events are simultaneous? Does there exist a spaceship frame in which the two events occur at the same point of space? In such a spaceship frame, what is the spatial distance or the time interval between the two events?

21. The speed of transverse waves on a wire is $v = \sqrt{T/\mu}$, where T is the tension and μ the mass per unit length.
(a) Since the maximum speed for such a wave (or any other signal) is c, what can you conclude about the maximum conceivable tension that can exist in a wire?
(b) Experimental measurements indicate that the actual maximum tension that a steel wire of radius 1 mm can support without breaking is 1.6×10^3 N. How does this compare with the relativistic limit found in part (a)? The density of steel is 7.8×10^3 kg/m^3.

22. Consider two speeds V_1 and V_2, each of which is less than the speed of light. Show that if these speeds are combined by the relativistic combination formula $(V_1 + V_2)/(1 + V_1 V_2/c^2)$, the result is always less than the speed of light.

23. Some quasars, such as 3C345, have been observed to eject luminous blobs of gas that appear to move away from the quasar with a transverse speed in excess of the speed of light. The explanation of this illusion is that the blob is actually moving at a speed slightly less than the speed of light in a direction that makes a small angle with the line of sight (see Fig. 2.25). The light emitted when the blob is born at Q and the light emitted when the blob arrives at P then reach the Earth

almost simultaneously, creating the illusion that the blob moves almost instantaneously from Q to P.
(a) Suppose that the time interval between the birth of the blob at Q and its arrival at P is Δt_0. Show that if $\theta \ll 1$, the time interval between reception at the Earth of the light emitted at Q and reception of the light emitted at P is $\Delta t = (1 - v/c)\Delta t_0$, where v is the speed of the blob.
(b) Show that the apparent transverse speed is

$$\frac{PP'}{\Delta t} \simeq \frac{\theta' OP}{(1 - v/c)(QP/v)}$$

(c) Assume that $OP/QP \simeq 100$ and $\theta' \simeq 1$ arcsec. For what values of v will the apparent transverse speed $PP'/\Delta t$ exceed c?

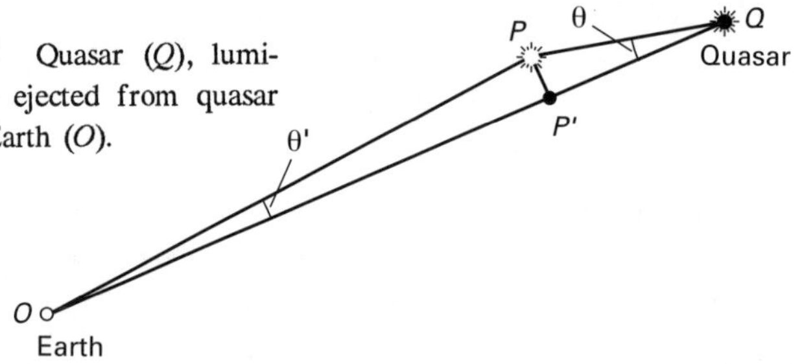

Fig. 2.25 Quasar (Q), luminous blob ejected from quasar (P), and Earth (O).

24. Derive the Lorentz transformation equations (38) and (39) for the y and z components of the velocity.

25. In the spaceship frame, a particle moves with speed v' along a straight line in the x'-y' plane; the direction of motion of the particle makes an angle θ' with the x' axis.
(a) Find the x' and y' components of the velocity in the reference frame of the spaceship. Find the x and y components of the velocity in the reference frame of the laboratory (as always, the spaceship moves with speed V along the x axis of the laboratory frame).
(b) Show that in the laboratory frame, the direction of motion of the particle makes an angle θ with the x axis such that

$$\tan\theta = \frac{v'\sqrt{1 - V^2/c^2}\,\sin\theta'}{v'\cos\theta' + V}$$

(c) Suppose that a source at the origin of the spaceship frame emits particles in all directions. Show that, for $V \rightarrow c$, all these particles, except those emitted exactly in the backward direction ($\theta' = 180^0$), will form a narrow beam in the forward direction ($\theta = 0$) in the laboratory frame.

26. A spaceship launches a three-stage rocket which attains a speed v_b when the first stage burns out. The second stage then separates from the first and, at its burnout, attains a forward speed of v_b relative to the freely coasting first stage. Finally, the third stage separates from the second and, at its burnout, attains a forward speed of v_b relative to the freely coasting second stage. What is the final speed of the third stage relative to the spaceship?

27. In the laboratory frame, particle 1 moves along the x axis with a uniform velocity u_x, and particle 2 moves along the y axis with a uniform velocity v_y. Find the velocity of particle 2 in the rest frame of particle 1. Find the velocity of particle 1 in the rest frame of particle 2.

28. In a transparent medium of index of refraction n, the speed of light relative to the medium is c/n. Use the relativistic velocity combination formula to show that if the medium has a speed V relative to the laboratory, then the speed of light relative to the laboratory is approximately given by the formula

$$c' = \frac{c}{n} \pm \left(1 - \frac{1}{n^2}\right)V$$

where the + sign is valid when $\mathbf{V}$ is parallel to $\mathbf{c}$, and the - sign is valid when $\mathbf{V}$ is antiparallel to $\mathbf{c}$. (As mentioned in Chapter 1, this formula was first obtained by Fresnel, who thought that the moving medium drags the ether along, giving it a drag speed of $(1 - 1/n^2)V$. Fresnel assumed that the speed of light was c/n *relative to the ether*, and he obtained c' by adding this speed and the drag speed, according to the Galilean velocity addition formula.)

29. In the laboratory frame, a light pulse travels in the x-y plane, with $v_x = c \cos \phi$ and $v_y = c \sin \phi$. Find the components v'_x and v'_y of the velocity of this light pulse in the spaceship frame. Evaluate $v'^2_x + v'^2_y$, and verify explicitly that it equals c^2.

30. Aberration is the change in the direction of propagation of starlight caused by the motion of the Earth around the Sun. The simplest instance of aberration occurs for stars located perpendicular to the plane of the Earth's orbit (directly "above" the orbit). In the reference frame x',y',z' centered on the Sun, the light from such a star propagates downward along the y' axis, perpendicular to the plane of the Earth's orbit, with velocity components $v'_x = 0$, $v'_y = -c$, $v'_z = 0$. Find the components of the velocity of the light in the reference frame centered on the Earth. Show that in this reference frame, the direction of propagation of the light makes angle

$$\phi = \tan^{-1}(\gamma V/c)$$

with the y axis. For $V = 30$ km/s. What is the value of ϕ in arcsec?

31. In the rest frame of a (flat) mirror, the law of reflection states that the angle of reflection of a light ray equals the angle of incidence.

(a) Suppose that the mirror is moving relative to the laboratory with a velocity V

in a direction perpendicular to its face. If the angle of incidence of a light ray is θ (measured in the laboratory), what is the angle of reflection?
(b) Suppose that the mirror is moving relative to the laboratory with a velocity V in a direction parallel to its face. If the angle of incidence of a light ray is θ (measured in the laboratory), what is the angle of reflection?

32. Derive the Lorentz transformation equations (42) and (43) for the y and z components of the acceleration.

33. In the laboratory frame, a particle is performing uniform circular motion in the x-y plane, according to the equations

$$x = A \cos \omega t$$

$$y = A \sin \omega t$$

The acceleration of this particle is centripetal, of magnitude $\omega^2 A$. According to Eqs. (41) and (42), what are the components of the acceleration of this particle in the spaceship frame at the spacetime point $t = 0, x = A, y = 0$? at the spacetime point $t = \pi/2\omega, x = 0, y = A$? at the spacetime point $t = \pi/\omega, x = -A, y = 0$? Is the acceleration in the spaceship frame centripetal?

34. Consider a spaceship that accelerates according to Eq. (46), starting from rest. How long a time is required to achieve a speed of $0.5c$ if the acceleration in the spaceship frame is constant at $a'_x = g$? How far will the spaceship have traveled in this time?

35. (a) By eliminating the time from Eqs. (46) and (48), show that

$$x = \frac{c^2}{a'_x}\left(\frac{1}{\sqrt{1 - v_x^2/c^2}} - 1\right)$$

Show that for low velocities, this coincides approximately with the familiar Newtonian result $2a'_x x = v_x^2$.
(b) As stated in Section 2.6, for a charged particle in a uniform electric field, $a'_x = eE/m$. If we want to accelerate an electron in a uniform electric field and attain a speed of $0.8c$ within a distance of 20 cm, what electric field do we need?

3

Time Dilation and Length Contraction

AS FOR ME, I DO NOT BELIEVE IN MATHEMATICS.

Albert Einstein to Gustave Ferrière

The Lorentz transformations derived in the preceding chapter are a central feature of the theory of Special Relativity. All the physical effects predicted by the theory can be extracted from the Lorentz transformations. In this chapter we will investigate some of the most striking of these effects: the time dilation, the relativistic Doppler shift, and the length contraction. We will also present the experimental evidence for the time dilation and the Doppler shift. There is not yet any direct experimental evidence for the length contraction, although there is some indirect evidence from measurements of the lifetimes of fast-moving elementary particles and from the behavior of electric and magnetic forces.

3.1 TIME DILATION

Suppose that a clock is moving relative to the laboratory frame at a high speed V. We can regard this moving clock as belonging to a spaceship frame. We want to measure the ticks of this moving clock by means of the clocks belonging to the laboratory frame and thereby compare its rate with the rate of the laboratory clocks. We can regard each tick of the moving clock as an event. Figure 3.1 is a Minkowski diagram showing these events. For the sake of simplicity, assume that in the spaceship frame the moving clock ticks at intervals of $c\Delta t' = 1$, or $\Delta t' = 1/c = 1/(3 \times 10^8)$ s. Thus, the clock ticks at $ct' = 1$, $ct' = 2$, $ct' = 3$, etc. To find the laboratory time coordinates corresponding to these ticks, we draw horizontal lines from the tick events to the ct axis. As can be seen in Fig. 3.1, these horizontal lines intercept the ct

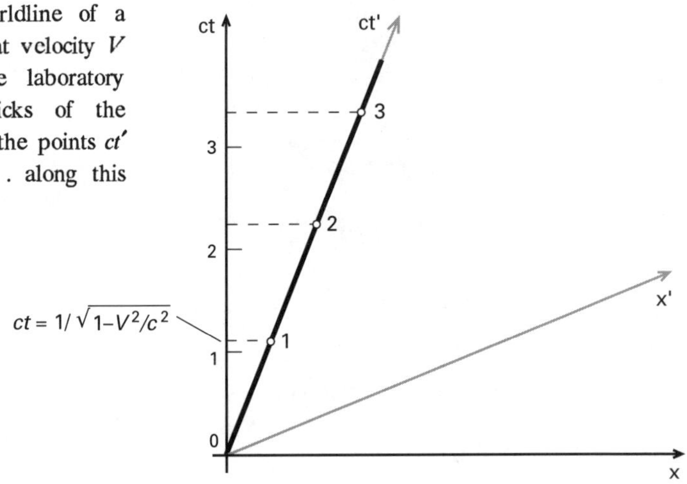

Fig. 3.1 Worldline of a clock moving at velocity V relative to the laboratory frame. The ticks of the clock occur at the points $ct' = 1, 2, 3, \ldots$ along this worldline.

axis at intervals exceeding $c\Delta t = 1$. Accordingly, as measured by the laboratory clocks, the spaceship clock runs slow. From our discussion of the calibration of the ct' axis in Section 2.4, we know that the horizontal line from the unit point $ct' = 1$ intercepts ct axis at $ct = 1/\sqrt{1 - V^2/c^2}$. Hence we conclude that relative to the laboratory clocks, the spaceship clock runs slow by a factor $\sqrt{1 - V^2/c^2}$. This is the **time dilation** effect of special relativity.

We can also calculate the time dilation algebraically from the Lorentz transformation equations. From Eq. (2.17) we find that the space and time displacements between one tick of the clock and the next are related by

$$\Delta t = \frac{1}{\sqrt{1 - V^2/c^2}}(\Delta t' + V\Delta x'/c^2) \tag{1}$$

Since the spaceship clock is at rest in the spaceship reference frame, its x' coordinate is fixed, so $\Delta x' = 0$ and therefore

$$\boxed{\Delta t = \frac{1}{\sqrt{1 - V^2/c^2}}\Delta t'} \quad \text{(clock at rest in spaceship, } \Delta x' = 0\text{)} \tag{2}$$

This shows that the rate of the spaceship clock is slow by a factor $\sqrt{1 - V^2/c^2}$, in agreement with the argument given above.

Another way to obtain the time dilation is by exploiting the invariance of the spacetime interval between one tick of the spaceship clock and the next:

$$c^2\Delta t^2 - \Delta x^2 = \Delta t'^2 - \Delta x'^2 \tag{3}$$

In the spaceship frame, the clock is at rest, so $\Delta x' = 0$. In the laboratory frame, the clock moves at speed V, so $\Delta x = V\Delta t$. With these substitutions, Eq. (3) becomes

$$c^2\Delta t^2 - V^2\Delta t^2 = \Delta t'^2 \tag{4}$$

This equation is equivalent to Eq. (2).

Yet another way to obtain the time dilation is by a detailed examination of the operation of a clock. For this purpose, it is best to use a clock with an extremely simple mechanism so that the effect of motion on the operation of the clock can be analyzed easily. The simplest clock consists of two parallel mirrors held at a fixed distance, with a pulse of light bouncing back and forth between the mirrors. Each back-and-forth motion of the light pulse counts as one tick of the clock (see Fig. 3.2). We will suppose that the distance L between the mirrors is at right angles to the direction of motion, so that the length contraction does not affect this distance. The mirrors are at rest in the spaceship frame, and in this frame the time for one tick of the clock, or the round-trip time for the light pulse, is

$$\Delta t' = 2L/c \tag{5}$$

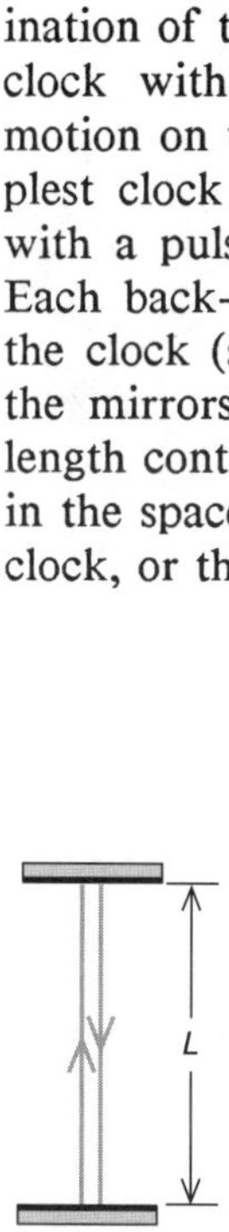

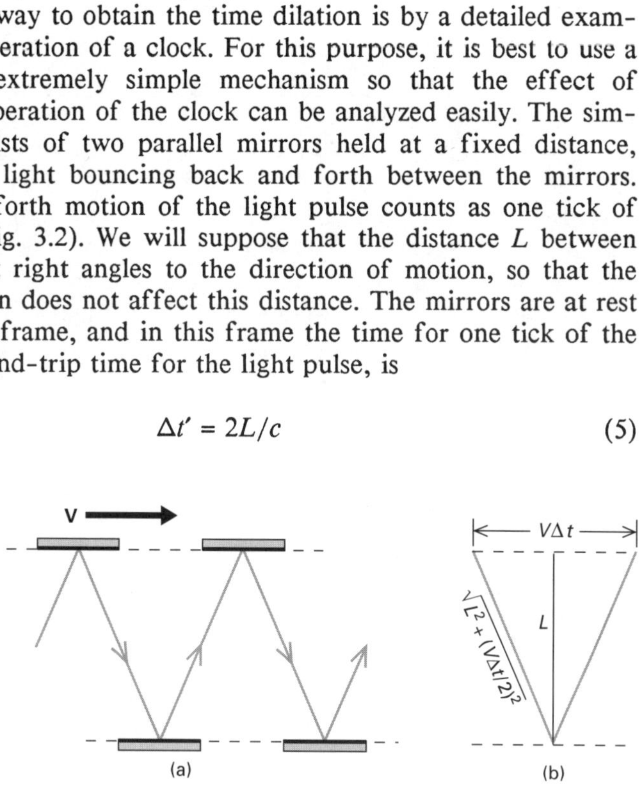

Fig. 3.2 (Left) Two mirrors with a light signal bouncing back and forth.

Fig. 3.3 (Right) (a) The mirrors and the light signal are moving toward the right relative to the laboratory frame. (b) The downward and upward paths of the light pulse are hypotenuses of right triangles.

In the laboratory frame, the mirrors travel toward the right at speed V; correspondingly, the motion of the light pulse has both an up or down component and a horizontal component (see Fig. 3.3a). Thus, in the laboratory frame, the length of path for a round trip of the light pulse is longer, and since the speed of light is c in both reference frames, the travel time of the light pulse in the laboratory frame is longer. The upward path is the hypotenuse of a right triangle with sides L and $V\Delta t/2$ (see Fig. 3.3b), so this upward path has a length $\sqrt{L^2 + (V\Delta t/2)^2}$. The downward path has the same length, and the time for one tick of the clock is therefore

$$\Delta t = 2 \times \sqrt{L^2 + (V\Delta t/2)^2}\ /c \tag{6}$$

Solving this for Δt, we obtain

$$\Delta t = \frac{2L/c}{\sqrt{1 - V^2/c^2}} \tag{7}$$

Substituting $2L/c = \Delta t'$ according to Eq. (5), we find the same time dilation factor as above.

The time-dilation effect is symmetric: as measured by the spaceship clocks, each of the laboratory clocks runs slow by the same factor $\sqrt{1 - V^2/c^2}$. Figure 3.4 shows the ticks of the laboratory clock in a Minkowski diagram, and it shows the slanted lines that intercept the ct' axis and indicate the time intervals measured by the spaceship clocks. From the figure we see that these time intervals are, indeed, dilated. This time dilation for a laboratory clock in motion relative to the spaceship frame can also be calculated algebraically from the Lorentz transformation equation (2.15), with $\Delta x = 0$:

$$\Delta t' = \frac{1}{\sqrt{1 - V^2/c^2}}\,\Delta t \quad \text{(clock at rest in laboratory, } \Delta x = 0) \tag{8}$$

We can summarize both Eq. (2) and Eq. (8) by introducing the concept of **proper time**, that is, the time registered by a clock in its own reference frame (the time shown on the clock's dial). If $\Delta\tau$ repre-

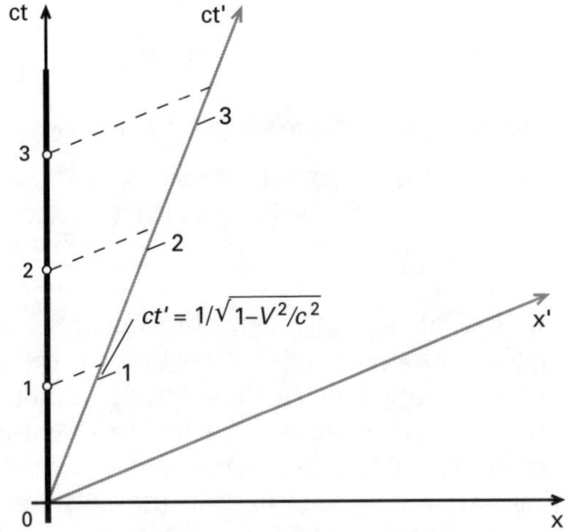

Fig. 3.4 Worldline of a clock at rest in the laboratory frame. The ticks of this clock occur at the points $ct = 1, 2, 3, \ldots$

sents the elapsed proper time and Δt the time measured in some inertial reference frame relative to which the clock is in motion, then the relation between $\Delta\tau$ and Δt is

$$\Delta\tau = \sqrt{1 - V^2/c^2}\,\Delta t \tag{9}$$

We can express the right side of this equation as follows:

$$\sqrt{1 - \frac{V^2}{c^2}}\,\Delta t = \sqrt{1 - \left(\frac{\Delta x}{c\Delta t}\right)^2}\,\Delta t = \frac{1}{c}\sqrt{(c\Delta t)^2 - (\Delta x)^2} = \frac{\Delta s}{c} \tag{10}$$

from which we see that $\Delta\tau$ equals the spacetime interval along the worldline of the clock, divided by c. But note that $\Delta\tau$ is defined only for timelike worldlines and is always a positive quantity, whereas, in general, the spacetime interval Δs is also defined for spacelike worldlines, in which case it is imaginary.

We can write Eq. (9) in differential form,

$$\boxed{d\tau = \sqrt{1 - V^2/c^2}\,dt} \tag{11}$$

This has the advantage that it can be applied to a clock with accelerated motion, with an instantaneous speed V that varies with time. Of course, the application of Eq. (11) to an accelerated clock hinges on the implicit assumption that the clock is constructed in such a way that it is not directly sensitive to accelerations.

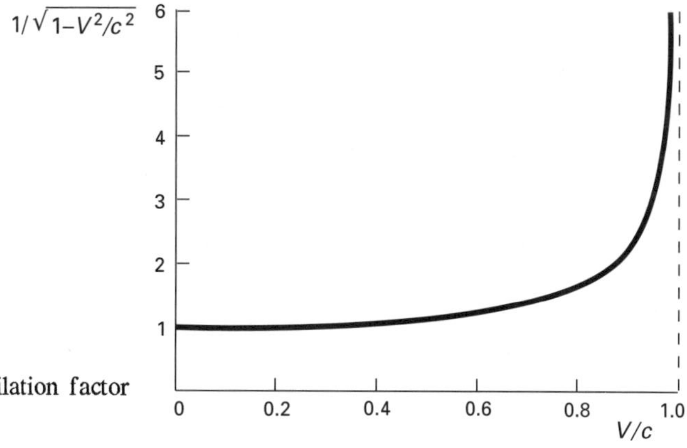

Fig. 3.5 The time-dilation factor $1/\sqrt{1 - V^2/c^2}$ vs. V.

Figure 3.5 is a plot of the time-dilation factor $1/\sqrt{1 - V^2/c^2}$ as a function of V. The time dilation becomes very large when $V \simeq c$, but it is insignificant when $V << c$. For low speeds, we can approximate the time-dilation factor as

$$\frac{1}{\sqrt{1 - V^2/c^2}} \simeq 1 + \frac{1}{2} V^2/c^2 \tag{12}$$

■ **EXAMPLE 1.** An astronaut in a spacecraft is orbiting the Earth in a low-altitude orbit at a speed of 6.8×10^3 m/s relative to the reference frame centered on the Earth. What is the time dilation for a clock in the spacecraft relative to the clocks of the reference frame centered on the Earth?

SOLUTION. With $V = 6.8 \times 10^3$ m/s, the approximation (12) gives a time-dilation factor of

$$\frac{1}{\sqrt{1 - V^2/c^2}} \simeq 1 + \frac{1}{2} V^2/c^2 = 1 + \frac{1}{2} \frac{(6.8 \times 10^3 \text{ m/s})^2}{(3.00 \times 10^8 \text{ m/s})^2} = 1 + 2.6 \times 10^{-10}$$

Thus, the clock of the spacecraft runs slow by a few parts in 10^{10}. Note that this result is not directly applicable to a comparison between the clock of the spacecraft and clocks at rest on the ground. Because the latter clocks participate in the rotation of the Earth, they are not in an inertial reference frame, and they have a speed and a time dilation relative to the (nonrotating) inertial reference frame centered on the Earth. However, even at the equator, the speed of rotation is less than a tenth of the orbital speed of the spacecraft, and therefore this rotational speed does not alter the time dilation by very much. In most calculations of time dilation, the rotation of the Earth can be neglected, and reference frames on the surface of the Earth can be treated as *approximately* inertial. Only for extremely precise tests of the time dilation do we have to take the rotation of the Earth into account (in Section 3.3, we will deal with an example of a precise test in which the rotational speed plays a role).

It is important to understand that the time dilation applies to all kinds of clocks and to all kinds of processes involving physical changes with a definite rate in time. Thus, in Example 1, the clock of the spacecraft suffers time dilation, and so does the wristwatch of the astronaut, and so does the biological body clock of the astronaut--she ages more slowly than her colleagues on Earth. At the speed assumed in Example 1, the time-dilation effect is too small to produce a noticeable difference in age or a noticeable difference in the time shown on a wristwatch. However, accurate atomic clocks can detect small discrepancies in time, and they have been used in experimental tests of the time dilation (see Section 3.3).

Note that the relativistic time dilation spoils the Newtonian procedure for synchronization of clocks by means of clock transport. The transported clock is in motion relative to the clocks of the inertial reference frame; it therefore runs slow, and it cannot be used to transfer the synchronization from one location to another. However, synchronization by clock transport remains valid if we proceed to the limit of a very low speed of the transported clock. In this case, the deviation between the transported clock and the synchronized clocks of the reference frame remains small even if the transport distance is large. To establish this, assume that the transported clock has a constant speed V and so accumulates a proper time $\Delta\tau = \sqrt{1 - V^2/c^2}\,\Delta t$ during the transport. It therefore deviates from the synchronized clocks of the reference frame by

$$\Delta\tau - \Delta t = \sqrt{1 - V^2/c^2}\,\Delta t - \Delta t \simeq -\frac{1}{2}\frac{V^2}{c^2}\,\Delta t \tag{13}$$

Since $V\Delta t$ equals the transport distance L, we can write this deviation as

$$\Delta\tau - \Delta t = -VL/2c^2 \tag{14}$$

This approaches zero for $V \to 0$, regardless of how large the transport distance is. From this we see that instead of synchronizing the clocks of the reference frame by light signals, according to the procedure described in Section 2.2, we could equally well synchronize them by clock transport, provided that we used transported clocks moving at very low speeds.

Time dilation plays a significant practical role in the operation of the Global Positioning System (GPS). The accurate determination of latitude and longitude by means of time signals sent from the GPS satellites to a portable GPS receiver requires that the clocks on board these satellites be synchronized to within a few nanoseconds. To maintain this synchronization, the clocks must be corrected for time dilation.

Time dilation also plays a practical role in the operation of radiotelescopes with the VLBI technique mentioned in Section 1.7. The clocks used to "tag" the recordings of the radio signals received at two or more radiotelescopes separated by a few thousand kilometers must be synchronized to within a nanosecond. To achieve such an accurate synchronization, the clocks must be corrected for time dilation, and they must also be corrected for the relativity of synchronization (the term Vx/c^2 discussed in the Section 2.3).

Incidentally, the operation of the GPS relies on relativity not only for the corrections for time dilation, but also for the calculation of the latitude and longitude from the travel times of the radio signals, which hinges on the assumption of a fixed, direction-independent speed of light. The same is true for the operation of radiotelescopes with the VLBI technique.

3.2 THE DOPPLER SHIFT OF LIGHT

It is well known that for sound waves in air, whenever the emitter of waves is in motion relative to the receiver of waves, the received sound wave is subjected to a frequency shift called the **Doppler shift**.

Light waves are subjected to a similar Doppler shift. However, in contrast to sound waves in air, light waves in vacuum are not propagating in any material medium that can serve as a preferred reference frame with respect to which we can reckon their speed. Light waves have the same speed c with respect to all inertial reference frames, and therefore we can use any reference frame whatsoever for calculating the Doppler shift of light.

In the following calculation we will adopt the reference frame of the emitter. Suppose that in this reference frame, the light wave has a frequency ν and a wavelength $\lambda = c/\nu$. Figure 3.6 shows the worldlines of successive wavecrests of the light wave. The horizontal distance between the wavecrests is the wavelength λ, and the vertical distance between them is the period multiplied by c, or $1/\nu \times c$. The figure also shows the worldline of the receiver, which is moving away from

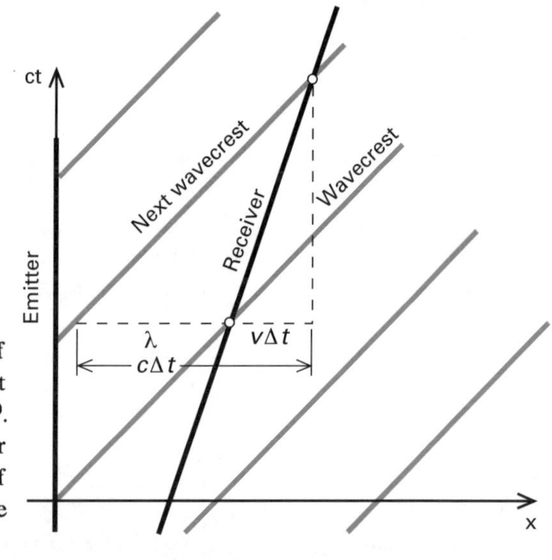

Fig. 3.6 The worldlines of the wavecrests of the light wave slope upward at 45°. The worldline of the receiver intersects the worldline of one wavecrest and then the next.

the emitter at speed v in the positive x direction. Detection of wavecrests by the receiver occurs at the intersections of the wavecrest worldlines with the receiver worldline. From the figure we see that after one wavecrest is detected by the receiver, the distance the next wavecrest has to travel to the receiver is the sum of wavelength λ and the distance $v\Delta t$ traveled by the receiver,

$$\Delta x = \lambda + v\Delta t$$

and this must equal the distance $c\Delta t$ traveled by the wavecrest in the same time,

$$c\Delta t = \lambda + v\Delta t$$

Thus, the time interval between detection of one wavecrest and detection of the next is

$$\Delta t = \frac{\lambda/c}{1 - v/c} = \frac{1}{\nu}\,\frac{1}{1 - v/c} \tag{15}$$

This time interval is measured in the reference frame of the emitter. But during the emitter time interval Δt, the receiver's clock registers a shorter proper time interval $\Delta t' = \Delta t\sqrt{1 - v^2/c^2}$. Hence Eq. (15) becomes

$$\frac{\Delta t'}{\sqrt{1 - v^2/c^2}} = \frac{1}{\nu}\,\frac{1}{1 - v/c} \tag{16}$$

For the frequency $\nu' = 1/\Delta t'$ detected by the receiver, we therefore find

$$\nu' = \nu\,\frac{1 - v/c}{\sqrt{1 - v^2/c^2}} \tag{17}$$

With $1 - (v/c)^2 = (1 - v/c)(1 + v/c)$, we can simplify this to obtain

$$\boxed{\nu' = \sqrt{\frac{1 - v/c}{1 + v/c}}\;\nu} \qquad \text{(for receding receiver and emitter)} \tag{18}$$

If the receiver is moving toward the emitter, the minus sign in Eq. (16) must be replaced by a + sign, and we obtain

$$\boxed{\nu' = \sqrt{\frac{1 + v/c}{1 - v/c}}\;\nu} \qquad \text{(for approaching receiver and emitter)} \tag{19}$$

These are the relativistic formulas for the Doppler shift of light.

Note that the factor $(1 - v/c)$ in Eq. (17) is the familiar Doppler shift factor of Newtonian physics (for a receding receiver). The extra factor $1/\sqrt{1 - v^2/c^2}$ is the contribution from the time dilation. Thus, we could have obtained the relativistic formula by simply starting with the Newtonian formula and making a correction for time dilation.

The relativistic Doppler formula differs from the Newtonian formula only by terms of second order and higher. If $v/c << 1$, we can use the approximation $\sqrt{1 - v^2/c^2} \simeq 1 - v^2/2c^2$ for the square root in Eq. (17), which gives

$$\nu' \simeq \frac{1 - v/c}{1 - v^2/2c^2}\,\nu \simeq (\,1 - v/c)(1 + v^2/2c^2)\,\nu \tag{20}$$

This means that the difference between the Newtonian and the relativistic formulas is extremely small at the speeds that can be attained in laboratory experiments (for a discussion of experiments, see the next section).

The formulas (18) and (19) are valid for emitter and receiver receding or approaching in a radial direction. But relativity also predicts a Doppler shift when the emitter is moving across the line of sight from the receiver. This is called the **transverse** Doppler effect, in contrast to the **longitudinal** Doppler effect that occurs when the emitter is receding or approaching in the direction of the line of sight. In the transverse case, the Doppler shift is entirely due to the time dilation, which reduces the frequency of a moving source by the factor $\sqrt{1 - v^2/c^2}$. Thus, the formula for the transverse Doppler effect is

$$\nu' = \nu \sqrt{1 - v^2/c^2} \tag{21}$$

If $v << c$, we can again use the approximation for the square root, which leads to a frequency shift of

$$\nu' - \nu \simeq \nu\left(1 - \frac{1}{2}\frac{v^2}{c^2}\right) - \nu = -\frac{1}{2}\frac{v^2}{c^2}\nu \tag{22}$$

and a fractional frequency shift of

$$\frac{\nu' - \nu}{\nu} \simeq -\frac{1}{2}\frac{v^2}{c^2} \tag{23}$$

This transverse Doppler shift is of second order in v, in contrast to the longitudinal Doppler shift, which is of first order in v.

More generally, the velocity of the emitter relative to the receiver can have components in both the radial (longitudinal) direction and the transverse direction. If the velocity of the emitter makes an angle θ with the line of sight (see Fig. 3.7), then the general Doppler shift formula is

$$\nu' = \frac{\nu\sqrt{1 - v^2/c^2}}{1 - v/c \cos\theta} \tag{24}$$

This general formula coincides with (21) if $\theta = 90^0$, and it coincides with (18) and with (19) for $\theta = 180^0$ and 0^0, respectively.

Note that the relativistic Doppler shift formula for receding emitter and receiver has an interesting multiplicative property: Suppose that the receiver has a speed v relative to the emitter, and a second receiver has a speed V relative to the first receiver, from

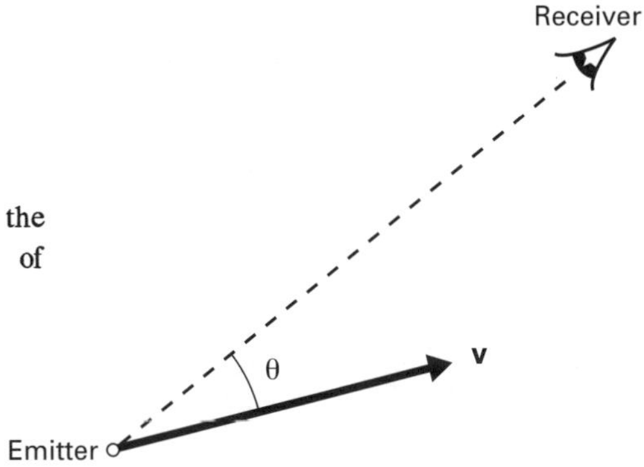

Fig. 3.7 The angle θ between the line of sight and the direction of motion of the emitter.

which it receives a retransmitted light signal. Then the net Doppler shift factor between the emitter and the second receiver is given by the product of two Doppler shift factors:

$$\sqrt{\frac{1 - v/c}{1 + v/c}} \times \sqrt{\frac{1 - V/c}{1 + V/c}} \tag{25}$$

Alternatively, this net Doppler shift factor can be expressed in terms of the speed of the emitter relative to the second receiver:

$$\sqrt{\frac{1 - v'/c}{1 + v'/c}} \tag{26}$$

From the velocity combination formula, we know that $v' = (v + V)/(1 + vV/c^2)$. Substituting this into (26), we find that (25) and (26) coincide. Thus, the velocity combination formula is consistent with the multiplicative property of the Doppler shift formula. We could have used this to derive the Doppler shift formula from the velocity combination formula, or vice versa.

■ **EXAMPLE 2.** Suppose that you are racing toward a red traffic light in a fast spaceship. At what speed will the red light look green to you?

SOLUTION. Red light has a wavelength of 650 nm, and green light has a wavelength of 500 nm. These wavelengths are inversely proportional to the emitted and the received frequencies; hence $\nu'/\nu = 650/500 = 1.3$. According to Eq. (19), the ratio ν'/ν is related to v/c:

$$\sqrt{\frac{1 + v/c}{1 - v/c}} = \frac{\nu'}{\nu}$$

Solving this for v/c, we find

$$\frac{v}{c} = \frac{(\nu'/\nu)^2 - 1}{(\nu'/\nu)^2 + 1} = \frac{(1.3)^2 - 1}{(1.3)^2 + 1} = 0.26$$

3.3 EXPERIMENTS ON TIME DILATION AND DOPPLER SHIFT

The slowing down of the rate at which time elapses applies to all physical processes: atomic, nuclear, biological, etc. When the rates of such processes are reckoned in proper time, they are independent of the speed of the atom, nucleus, or biological organism relative to the laboratory. But when the rates are reckoned in laboratory time, they are slowed down by the time-dilation factor. At low speeds this slowing is insignificant, but at high speeds it can become quite appreciable. For instance, the laboratory lifetimes of elementary particles are drastically increased when these particles move at high speed relative to the laboratory. One of the first experiments testing this effect was performed with muons, which are unstable particles with an average lifetime of only 2.2×10^{-6} s when at rest in the laboratory. But if such a muon is moving through the laboratory at a speed of $0.995c$, its lifetime is dilated by a factor

$$\frac{1}{\sqrt{1 - V^2/c^2}} = \frac{1}{\sqrt{1 - (0.995)^2}} = 10.01 \tag{27}$$

Quantitative measurements of the lifetime of muons as a function of speed were first performed by Rossi and Hall in 1941 [2]. They verified an increase of lifetime of the order of magnitude indicated by (27) in measurements on muons produced by the impact of primary cosmic rays on the top of the atmosphere. These muons travel 10 or 20 km from the top of the atmosphere to the ground. Without time dilation, hardly any of them would survive the trip to the ground, since a muon with the maximum speed 3×10^8 m/s, would be able to travel an average maximum distance of only $3 \times 10^8 \text{ m/s} \times 2.2 \times 10^{-6} \text{ s} \simeq 660$ m. Thus, the abundant flux of cosmic-ray muons found at ground level is a direct consequence of time dilation. A more recent version of the Rossi-Hall experiment confirmed the relativistic formula for time dilation to within the experimental uncertainty of 4% [7].

The most precise measurements of the increase of lifetime of elementary particles have been performed with muons and with pions produced by accelerators [9, 11]. The experimental uncertainties for the best of these measurements were of the order of 0.1%.

As we saw in Section 3.2, the time dilation gives rise to a transverse Doppler shift, that is, a frequency shift of the light emitted by a moving source when the line of sight is at right angles to the direction of motion of the source. This transverse Doppler shift has been measured in the light emitted by a beam of hydrogen atoms with $v = 3 \times 10^{-3}\, c$. For this speed, the fractional frequency shift (23) is

$$\frac{\nu' - \nu}{\nu} = -4 \times 10^{-6} \tag{28}$$

Because the effect is so small, the direct detection of the second-order Doppler shift in the light received at right angles to the beam is not feasible; even a very slight deviation from exact perpendicularity would introduce a much larger additional first-order Doppler shift, which would completely overwhelm the small second-order Doppler shift we seek to detect. This difficulty is bypassed with an ingenious method first proposed by Ives and Stilwell [1], who measured the *average* of the Doppler shifts for the light emitted in the forward and the backward directions along the beam of atoms. From Eq. (17) and the corresponding equation with opposite velocity, we see that this average Doppler shift factor is proportional to $1/\sqrt{1 - v^2/c^2}$, in contrast to the Newtonian average Doppler shift factor, which is zero. The best version of this experiment confirmed the relativistic Doppler shift to within 5% [6].

More precise tests of the relativistic time dilation have been performed by laser spectroscopy, in which a sample of gas is exposed to the beam from a tunable laser, and the frequency of the latter is adjusted to achieve resonant absorption of the laser light by the atoms. This resonance condition indicates that the frequencies are exactly matched and permits an extremely precise determination of the frequency of the atomic transition. Snyder and Hall [10] used such laser spectroscopy to compare the resonant frequency of neon atoms in a discharge cell with the resonant frequency of neon atoms in a fast atomic beam crossing the laser beam at right angles. With atoms of a speed of about $10^{-3}c$, they verified the time dilation to within about 0.5%. Juncar et al. obtained similar results using a laser beam parallel to a fast atomic beam [13]. MacArthur et al. verified the time dilation to within 0.03% by means of a laser beam intersecting a very fast atomic hydrogen beam at various angles [15]. The high speed of the atomic beam ($0.84c$) was attained by accelerating H^- ions with the LAMPF accelerator and then removing the surplus electron from these ions.

Kaivola et al. [14] used a two-laser technique, in which atoms of a fast atomic beam resonantly absorb *two* photons in succession from two oppositely directed laser beams while completing a single transition. One of these photons is supplied by a laser beam aimed in a direction parallel to the atomic beam, and the other photon is supplied by a laser beam aimed in the opposite direction, antiparallel to the atomic beam. Thus, the average Doppler shift factor of these photons is $1/\sqrt{1 - v^2/c^2}$, as in the Ives-Stilwell experiment. The experiment of

Kaivola et al. verified the time dilation to within 0.004%, and another version of this experiment, by McGowan et al. [17], verified the time dilation to within 0.0002%. Klein et al. [16] and Grieser et al. [18, 19] used an improved version of this two-laser technique with a beam of Li^+ ions of higher speed (0.064c), and they verified the time dilation to within 7×10^{-7}%. This is the most precise result that has been achieved so far.

Several tests of the relativistic time dilation have been performed by exploiting the Mössbauer effect (for details on the Mössbauer effect, see Section 5.5). Although the precision of these Mössbauer experiments has by now been surpassed by the laser experiments, the Mössbauer experiments remain of some interest because they provide direct empirical evidence that acceleration does not directly influence the time dilation.* In the experiments by Hay et al. [3] and by Kündig [8], the samples of ^{57}Fe that served as emitter and absorber of gamma rays were mounted at the center and at the rim, respectively, of an ultracentrifuge rotor. For a 15-cm rotor spinning at 35,000 rpm [8], the speed of the absorber is 5.5×10^2 m/s relative to the emitter. The frequency shift is simply the transverse Doppler shift given by Eq. (23). The predicted fractional frequency shift is

$$\frac{\nu' - \nu}{\nu} \simeq -\frac{1}{2}\left(\frac{5.5 \times 10^2 \text{ m/s}}{3 \times 10^8 \text{ m/s}}\right)^2 = -2 \times 10^{-12} \qquad (29)$$

Although this is quite a small frequency shift, it produces a reduction in the resonant absorption of the gamma rays. The most accurate version of this experiment has tested the time dilation to within 1.1%.

Another test of the time dilation is provided by the **thermal Mössbauer shift**. Suppose that the samples of ^{57}Fe that act as emitter and as absorber of gamma rays are both at rest in the laboratory. The individual atoms and nuclei will then still have a thermal motion, which produces a time dilation. We can calculate the time dilation from Eq. (11), which is valid even in the case of accelerated motion, provided that this accelerated motion is observed in an inertial reference frame, say, the laboratory frame. For a given time interval Δt, the elapsed proper time of a nucleus moving with speed V is

---

* Whether the acceleration of a clock affects its rate depends on the construction of the clock. Some clocks–for instance, pendulum clocks–are quite sensitive to acceleration, but this does not conflict with relativity, provided that we can explain the sensitivity to acceleration by applying the laws of (relativistic) mechanics to the machinery within the clock. The Mössbauer experiments give direct empirical evidence that the "machinery" within a nucleus is not sensitive to the acceleration attainable in ultracentrifuges, which is about 10^6 m/s^2.

$$\Delta\tau = \int_0^{\Delta t} \sqrt{1 - V^2/c^2}\, dt \simeq \int_0^{\Delta t} \left(1 - \frac{1}{2}\frac{V^2}{c^2}\right) dt \tag{30}$$

The integral cannot be evaluated directly, because the speed of the nucleus is time-dependent. Although we do not know the instantaneous speed of an individual nucleus, we know from thermal physics that the average of the square of the speed is directly proportional to the temperature and inversely proportional to the mass,

$$\overline{V^2} = \frac{3kT}{m} \tag{31}$$

Here T is the temperature, m the mass, and k is Boltzmann's constant. Hence the average proper time is

$$\Delta\tau = \left(1 - \frac{3}{2}\frac{kT}{mc^2}\right)\Delta t \tag{32}$$

Accordingly, we can regard $(1 - 3kT/2mc^2)$ as an average time-dilation factor for nuclei in thermal motion. This thermal time dilation reduces the frequency of vibration of the nucleus, as measured in the laboratory. If the temperatures of the emitting and the absorbing samples are T_s and T_a, respectively, then the frequency difference between the nuclei in these samples is

$$\nu_s - \nu_a = \left(1 - \frac{3}{2}\frac{kT_s}{m}\right)\nu - \left(1 - \frac{3}{2}\frac{kT_a}{m}\right)\nu = \frac{3}{2}\frac{k}{m}(T_a - T_s)\,\nu \tag{33}$$

and the fractional frequency difference is

$$\frac{\nu_s - \nu_a}{\nu} = \frac{3}{2}\frac{k}{m}(T_a - T_s) \tag{34}$$

For samples of ^{57}Fe, this fractional frequency difference amounts to $\simeq 10^{-15}$ per kelvin. The Mössbauer effect permits the detection of this small frequency shift. Measurements of the thermal Mössbauer shift by Pound and Rebka [4] confirmed the time dilation to within about 10%. And these measurements also confirmed that the time dilation does not depend (directly) on the acceleration. The nuclei in the samples of iron suffer violent accelerations of up to $10^{16} \times g$ in their thermal motions, but these accelerations have no detectable effect on their rate of vibration.

The time dilation has also been tested with atomic clocks. Häfele and Keating [5] carried atomic clocks on flights around the world on commercial aircraft and then compared them with clocks that remained on the ground. For the analysis of this experiment, we must take into account the rotation of the Earth. As seen from an inertial reference frame centered on the Earth (but not rotating with the Earth), clocks on the ground have an eastward speed; clocks flying eastward have a larger speed than clocks on the ground; and clocks flying westward have a smaller speed than clocks on the ground (see Fig. 3.8). Hence, as compared with clocks on the ground, clocks flying eastward should suffer a time dilation, and clocks flying westward should suffer a time "contraction." In agreement with this prediction, the experiment showed that the east-flying clocks lost time compared with the ground clocks and that the west-flying clocks gained time. The magnitude of the loss or the gain was of the order of 10^{-7} s for one complete trip around the Earth.

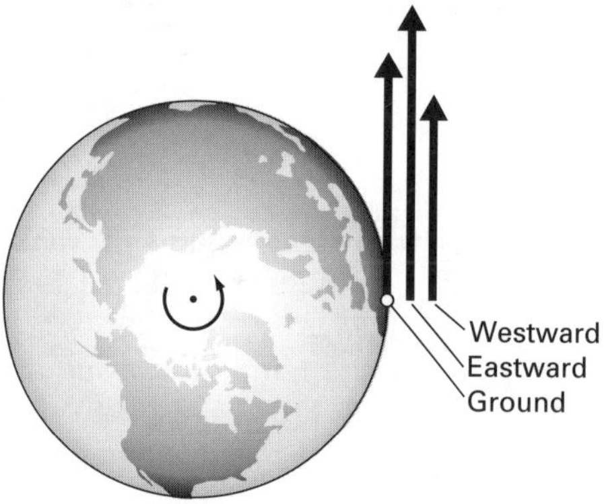

Fig. 3.8 Tangential speeds of the ground, an aircraft traveling eastward, and an aircraft traveling westward along the equator of the Earth.

The experimental measurements cannot be compared directly with the prediction of Special Relativity, because the clocks carried in the aircraft also suffer an extra gravitational effect. A clock on the ground is at a lower gravitational potential than a clock at high altitude, such as a clock at the typical cruising altitude of 10,000 m in a commercial aircraft. The clocks are then affected by the gravitional time-dilation effect of General Relativity, according to which the clock on the ground (at the lower gravitational potential) runs slower than the clock at high altitude. The experiment with traveling clocks was, in fact, intended primarily as a test of this general-relativistic effect. But the experiment also served as a test for the time-dilation effect of special relativity, because the east-flying and the west-flying clocks suffer the same gravitational effect (if at the same altitude); hence, the difference between these clocks must be attributed to their difference in speeds.

An accurate calculation of the predicted time dilation was made from the flight data of the aircraft. For a rough calculation of the expected magnitudes, let us assume that the aircraft flies along the equator, in the eastward direction. The proper time registered by the aircraft clock is then

$$\Delta\tau_1 = \int_0^t \sqrt{1 - V_1^2/c^2}\, dt \tag{35}$$

where V_1 is the speed of the aircraft as measured in an inertial frame. The proper time registered by the clock on the ground is

$$\Delta\tau_2 = \int_0^t \sqrt{1 - V_2^2/c^2}\, dt \tag{36}$$

Hence

$$\Delta\tau_2 - \Delta\tau_1 \simeq \int_0^t \left(- \frac{V_2^2}{2c^2} + \frac{V_1^2}{2c^2} \right) dt \simeq \int_0^t \frac{1}{2c^2} (V_1 - V_2)(V_1 + V_2) dt \tag{37}$$

$$\simeq \frac{1}{2c^2} (V_1 - V_2)(V_1 + V_2) t \tag{38}$$

where we have assumed that V_1 is approximately constant. Since $V_1 - V_2$ is the ground speed of the aircraft, $(V_1 - V_2)t$ is the distance traveled by the aircraft relative to the ground. Thus, for an eastward trip around the Earth,

$$\Delta\tau_2 - \Delta\tau_1 \simeq \frac{2\pi R(V_1 + V_2)}{2c^2} \tag{39}$$

where R is the radius of the Earth. With V_2 = (rotational speed at equator) $\simeq$ 460 m/s and V_1 = (aircraft speed) = (rotational speed) + (ground speed) $\simeq$ 460 m/s + 250 m/s, this equation yields a time loss of the order of $\simeq 2.6 \times 10^{-7}$ s for the aircraft clock. In the experiments, the time loss was measured to within 3%, and it agreed very well with the predicted time loss.

A more precise test of the time dilation with atomic clocks was performed by Vessot et al. [12]. In this test, a Scout rocket was used to launch several hydrogen-maser clocks into a nearly vertical trajectory, with a maximum speed of about 8 km/s and a maximum height of about 10,000 km. Because the clocks were destroyed when the rocket crashed at the end of its up-and-down trajectory, the clocks could not be compared with the ground clocks at the end of the flight. Instead they were compared during the flight, by means of radio signals

TABLE 3.1 TIME-DILATION EXPERIMENTS

Experimenters	Year	Fractional Experimental Uncertainty, %
Ives and Stilwell (Doppler shift) [1]	1938	3
Rossi and Hall (muon lifetime) [2]	1941	$\simeq$50
Hay et al. (Mossbauer, rotor) [3]	1960	4
Pound and Rebka (Mossbauer, thermal) [4]	1960	10
Hafele and Keating (atomic clocks) [5]	1962	3
Mandelberg and Witten (Doppler shift) [6]	1962	5
Frisch and Smith (muon lifetime) [7]	1963	4
Kundig (Mossbauer, rotor) [8]	1963	1
Ayres et al. (pion lifetime) [9]	1971	0.4
Snyder and Hall (Doppler shift, laser) [10]	1975	0.5
Bailey et al. (muon lifetime) [11]	1977	0.1
Vessot et al. (atomic clocks) [12]	1980	0.007
Juncar et al. (Doppler shift, laser) [13]	1985	0.5
Kaivola et al. (Doppler shift, two lasers) [14]	1985	0.004
MacArthur et al. (Doppler shift, laser) [15]	1986	0.03
Klein et al. (Doppler shift, two lasers) [16]	1992	0.001
McGowan et al. (Doppler shift, two lasers) [17	1993	0.0002
Grieser et al. (Doppler shift, two lasers) [18]	1994	7×10^{-7}

transmitted both ways between the ground station and the rocket. This experiment, like the flying clock experiment, was intended primarily to test the general relativistic time-dilation effect, but it also tested the special relativistic time-dilation effect. For a rocket coasting along a high-speed, high-altitude trajectory, the two effects are of the same order of magnitude; the special-relativistic effect dominates during the initial and final portions of the trajectory (at high speed), and the general-relativistic effect dominates when the rocket is near apogee (at high altitude). The rocket experiment confirmed the time dilation to within the experimental uncertainty of about 0.007%.

Table 3.1 summarizes the available experimental evidence for the relativistic time dilation.

3.4 LENGTH CONTRACTION

Consider a meter rod that is moving relative to the laboratory at speed V, in the direction of its length. In the rest frame of the rod, or the spaceship frame, at one instant of spaceship time, say $t' = 0$, the coordinates of the end of the rod are $x' = 0$, and $x' = 1$ m. Figure 3.9 is a Minkowski diagram with the worldlines of the front and the rear ends of this meter rod. The meter rod sweeps out the region of spacetime that is shown shaded in Fig. 3.9. How long is this meter rod as seen from the laboratory frame? At a given instant of laboratory time, say $t = 0$, the rear end of the meter rod is at A and the front end is at B. Since the point B is nearer the origin than the 1-meter calibration mark, the length is less than 1 meter as measured in the laboratory frame. From Section 2.4, we know that the slanted line $x' = 1$ m intercepts the x axis at $(1\text{ m}) \times \sqrt{1 - V^2/c^2}$. Thus, the length-contraction factor is $\sqrt{1 - V^2/c^2}$, and, in general, for a rod of length $\Delta x'$ at rest in the spaceship, the length measured at one instant of laboratory time is

$$\boxed{\Delta x = \sqrt{1 - V^2/c^2}\,\Delta x'} \qquad \text{(rod at rest in spaceship, measured at } t = 0\text{)} \qquad (40)$$

Figure 3.9 also gives us a qualitative insight into this contraction effect. When the laboratory observer determines the length of the rod by measuring the positions of its endpoints A and B at one instant of laboratory time, the observer in the spaceship reckons that the labora-

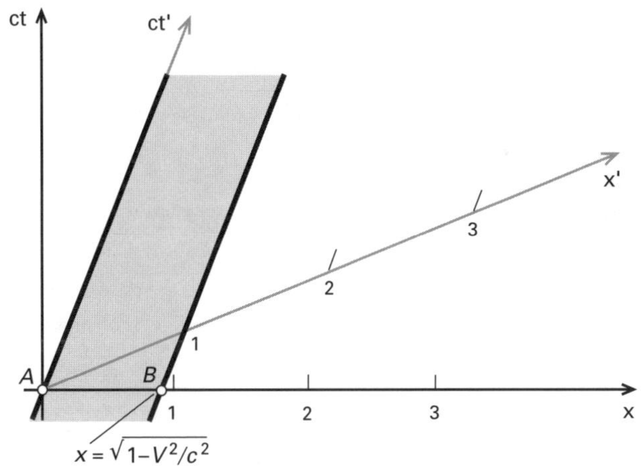

Fig. 3.9 Worldlines of the front and the rear of a meter rod in motion relative to the laboratory. The body of the meter rod sweeps out the region of spacetime shown shaded.

tory observer has measured the endpoints at unequal times. According to the judgment of the spaceship observer, the time of the B measurement is earlier than that of the A measurement. The spaceship observer will therefore attribute the short length obtained by the laboratory observer to a premature measurement of the leading end of the meter rod, when the leading end of the meter rod has not yet reached the "correct" position at a distance of 1 m from the trailing end. Thus, the length-contraction effect is a direct consequence of the relativity of synchronization.

For a meter rod at rest in the laboratory, the Minkowski diagram is as shown in Fig. 3.10. The measured length in the spaceship is AB, and this length is, again, contracted:

$$\Delta x' = \sqrt{1 - V^2/c^2}\,\Delta x \qquad \text{(rod at rest in laboratory, measured at } t' = 0) \tag{41}$$

The general conclusion is, then, that a moving body is always contracted in the direction of its motion. The length of a body in its own rest frame is called the **proper length**. This proper length is always longer than the length measured in any other reference frame.

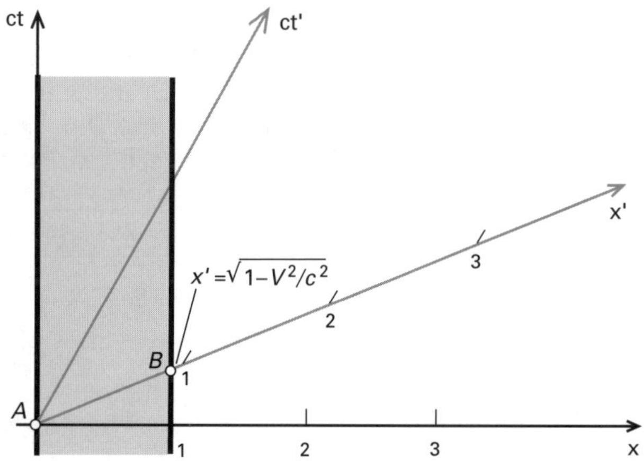

Fig. 3.10 Worldlines of the front and the rear of a meter rod at rest relative to the laboratory. The body of the meter rod sweeps out the region of spacetime shown shaded.

As in the case of the time dilation, we can also calculate the length contraction algebraically from the Lorentz transformation equations. According to Eq. (2.14), the space and time displacements are

$$\Delta x' = \frac{\Delta x - V\Delta t}{\sqrt{1 - V^2/c^2}} \tag{42}$$

For a rod at rest in the spaceship measured at one instant of laboratory time, we have $\Delta t = 0$, and therefore

$$\Delta x' = \frac{\Delta x}{\sqrt{1 - V^2/c^2}} \tag{43}$$

which is equivalent to the length-contraction formula (40). To obtain the length contraction for a rod at rest in the laboratory measured at one instant of spaceship time, we can start with the inverse Lorentz transformation equation (2.16).

The dimensions in the transverse directions remain unchanged. As a consequence, if a body has a volume $\Delta x'\Delta y'\Delta z'$ in its rest frame, then it will have a contracted volume

$$\Delta x\Delta y\Delta z = \Delta x'\Delta y'\Delta z' \sqrt{1 - V^2/c^2} \tag{44}$$

as measured in a reference frame with respect to which this body has speed V.

The length contraction alters the shape of a body (unless the body is two-dimensional and is placed at right angles to the direction of motion). For instance, a disk in the spaceship frame is distorted into an oval in the laboratory frame, and a square is distorted into a rectangle (see Figs. 3.11 and 3.12).

More severe distortions occur if the body is endowed not only with a high-speed translational motion but also with a high-speed rotational motion. For example, Fig. 3.13 shows the distortion of a wheel with spokes with a translational speed $V = 0.72c$ and a concurrent counterclockwise rotational speed $0.95c$, measured at the rim, in the rest frame of the center of the wheel. The distortion of the rotating wheel arises partially from length contraction and partially from the interplay of rotation and the relativity of synchronization. Opposite ends of a spoke reach, say, the horizontal position at one instant of time in the rest frame of the center of the wheel, or the spaceship frame. But these ends reach horizontal positions at unequal times in the laboratory frame (Leading Late!). Consequently, the straight spoke will not be straight in the laboratory frame. Note that in Fig. 3.13 the

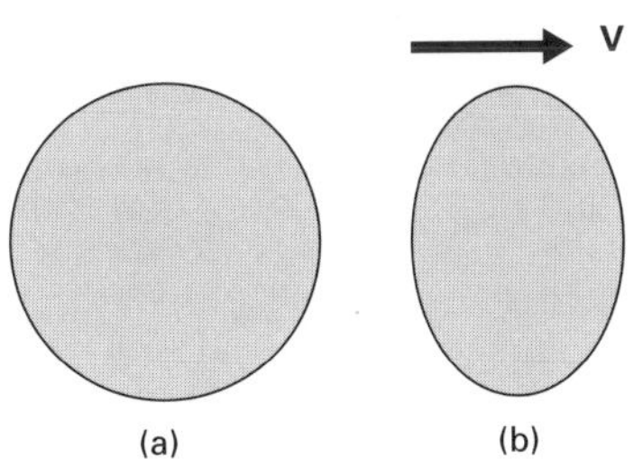

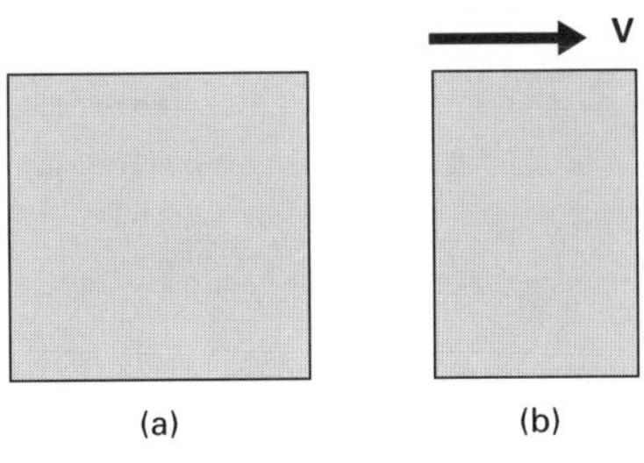

Fig. 3.11 (Left) Contraction of a disk oriented parallel to its direction of motion, with a translational speed $V = 0.72c$. (a) Configuration of the disk in its own rest frame, or spaceship frame. (b) Configuration of the disk at one instant of laboratory time.

Fig. 3.12 (Right) Contraction of a square oriented parallel to its direction of motion, with a translational speed $V = 0.72c$. (a) Configuration of the square in its own rest frame, or spaceship frame. (b) Configuration of the square at one instant of laboratory time.

spokes of the wheel are severely distorted, but the external boundary of the wheel has the same shape as the boundary of a nonrotating wheel (compare the wheel with the nonrotating disk in Fig. 3.11). This peculiarity of a rotating wheel arises from its rotational symmetry. In general, rotation leads to bizarre distortions of the external boundary. Figure 3.14 shows the distortion of a square with a translational speed $V = 0.72c$ and a concurrent rotational speed $0.95c$, measured at the vertices. (For more examples of bizarre distortions of bodies resulting from rotation or from acceleration, see Problems 35, 37, and 38.)

Although the relativistic length contraction is real, it would be difficult to observe directly, even if we had available a macroscopic body racing across our field of view at a high speed. If we were to photograph such a body, the contraction effect would be overwhelmed by time-delay effects. A photograph of a body does not show where the body is at a given time; rather, it shows the light that reaches the camera at a given time. This means that the light from those parts

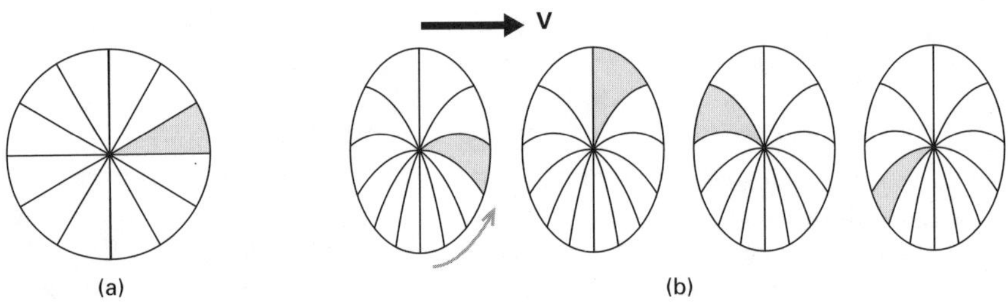

Fig. 3.13 Distortion of a wheel moving to the right with a translational speed $V = 0.72c$ and a concurrent counterclockwise rotational speed $0.95c$ (at the rim). (a) Configuration of the wheel at one instant of spaceship time. (b) Configuration of the wheel at several consecutive instants of laboratory time. The shaded sector helps to track the rotation of the wheel. Note that the deformation of each spoke changes as the wheel rotates.

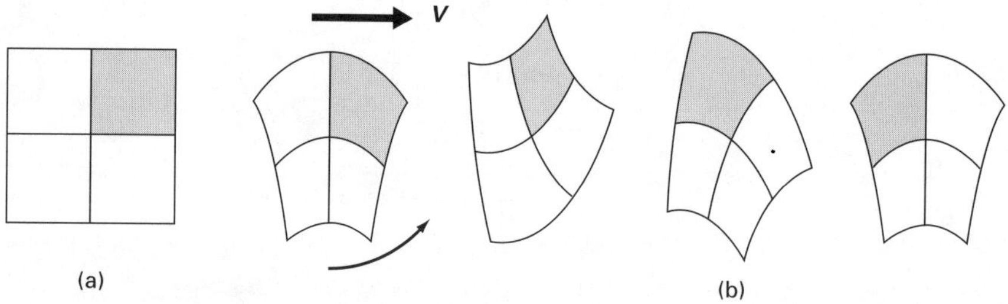

Fig. 3.14 Distortion of a square moving to the right with a translational speed $V = 0.72c$ and a concurrent counterclockwise rotational speed $0.95c$ (at the corners). (a) Configuration of the square at one instant of spaceship time. (b) Configuration of the square at several consecutive instants of laboratory time. The shaded quadrant helps to track the rotation of the square.

of the body that are farther away from the camera must be emitted earlier than the light from those parts that are nearer. Hence, quite apart the effects of the length contraction (and the additional effects of rotational motion, if present), the photograph of a fast-moving body will always exhibit some distortion.

Figure 3.15 shows what a photograph of a set of four parallelepipeds would look like [20]. Both the length contraction and the distortion arising from the time delays resulting from light propagation have been taken into account. The two parallelepipeds on the right show an obvious contraction, but the other two appear dilated. The combined effects of length contraction and time-delay distortion in (simulated) photographs have been calculated in detail for some other geometrical bodies. Although the effects are sometimes quite dramatic, they are usually dominated by the time-delay distortions. In fact, dramatic distortions of this kind would also appear if light were to propagate according to Newtonian physics rather than relativistic physics. Because of this, the time-delay distortions are not an essentially relativistic effect, and the insights about relativistic physics that can be extracted from such calculations of time-delay distortions are mostly insignificant.

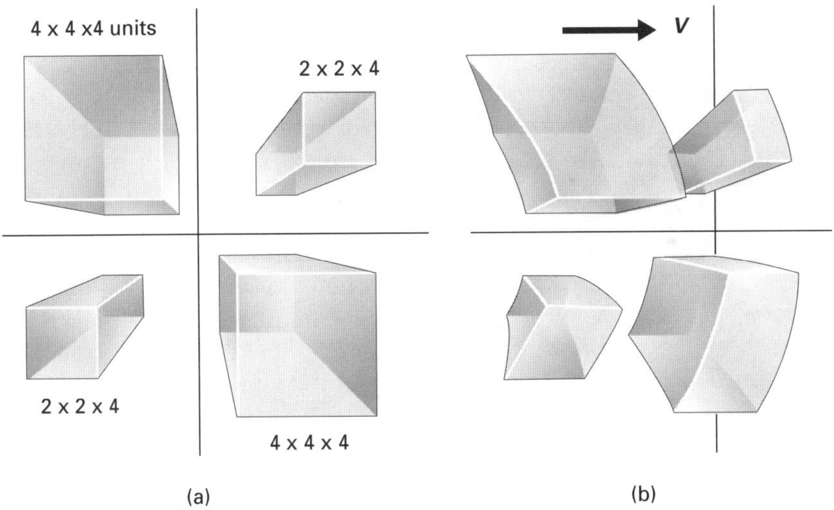

Fig. 3.15 Simulated photograph of a set of four parallelepipeds moving across the field of view at a speed of $V = 0.80c$. (a) Configuration in the rest frame, or spaceship frame. (b) Snapshot taken with a camera at rest in the laboratory. [After G. D. Scott and M. R. Viner, Am. J. Phys. **33**, 534 (1965)]

In the absence of direct observational evidence for the length contraction, we can rely on some indirect evidence. One such item of evidence emerges when we reconsider the case of cosmic-ray muons traveling downward through the atmosphere of the Earth. In Section 3.3 we examined the motion of the muons in the reference frame of the atmosphere, and we found that the muons survive the trip to the ground because of time dilation. But instead of examining the survival of the muon in the reference frame of the atmosphere, we can also examine it in the rest frame of the muon. In this rest frame, the atmosphere and the ground are rushing upward, past the stationary muon. Instead of attributing the survival of the muon to time dilation, we now have to attribute it to length contraction: in the rest frame of the muon, the atmosphere is contracted by a factor $\sqrt{1 - V^2/c^2}$, and it therefore takes less time for the ground to reach the stationary muon. Thus, the Rossi-Hall experiment and similar experiments on the dilation of lifetimes of elementary particles can be viewed as giving indirect evidence for the length contraction (the evidence is indirect since no actual measurements of length are performed). In Chapter 6 we will see that the behavior of electric and magnetic forces under transformations of the reference frame offers further indirect evidence for the length contraction.

3.5 THE TWIN PARADOX

The symmetric, mutual character of the time dilation gives rise to an apparent paradox, since it seems logically contradictory that the clocks in the spaceship reference frame run slow relative to the clocks in the laboratory frame, and concurrently the clocks of the laboratory frame run slow relative to those of the spaceship frame. This is called the **twin paradox**, because it leads to an apparent conflict in the final ages of two twins if one of the twins travels away from the other at high speed and then returns.

Suppose that Stella and Terra are identical twins, 20 years old. Stella leaves the Earth (regarded as an inertial "laboratory" frame) in a spaceship and travels at $V = \frac{1}{2}c$ to a neighboring star. She then reverses the motion of her spaceship (almost instantaneously) and returns to the Earth when Terra is 30 years old. How old is Stella at this time? The proper time (in years) elapsed along Stella's worldline is

$$\Delta\tau = \int_{20}^{30}\sqrt{1 - \frac{V^2}{c^2}}\,dt = \int_{20}^{30}\sqrt{1 - \left(\frac{1}{2}\right)^2}\,dt = \sqrt{\frac{3}{4}} \times 10 \simeq 8.7 \text{ years} \tag{45}$$

Hence, Stella is only 20 years + 8.7 years = 28.7 years old when the twins are reunited, that is, she is 1.3 years younger than Terra. However, as seen from the reference frame of Stella, Terra is moving at $V = \frac{1}{2}c$, and it would seem that Terra should suffer time dilation and that she should be the younger one. This is the twin paradox.

The resolution of this paradox hinges on the fact that Stella's reference frame is not inertial. To reverse her velocity and return to the Earth, Stella must accelerate. Although this acceleration has no effect on the aging of Stella as measured by Terra, it does have an effect on the aging of Terra as measured by Stella. Figure 3.16 shows the worldlines of Stella and of Terra in a Minkowski diagram. The coordinate axes x',ct' and x'',ct'' correspond to the rest frames of Stella for the outbound and the inbound portions of her voyage, respectively. Both of these reference frames are inertial. The rest frame for the outbound portion of the voyage has velocity V along the x axis of the x,t reference frame, and the rest frame for the inbound portion of the voyage has velocity $-V$. The origins of the coordinate axes x',ct' and x'',ct'' have been placed so the turnaround point P is at $t' = 0$, $t'' = 0$.

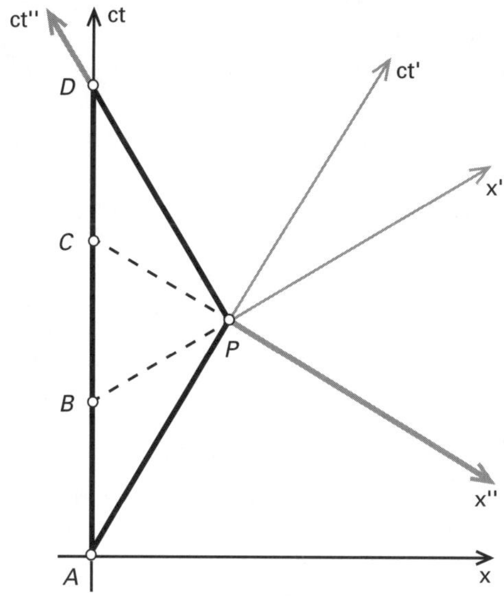

Fig. 3.16 Worldlines of Terra (AD) and of Stella (APD). The rest frame of Stella during the outbound portion of the trip has coordinate axes ct',x'. The rest frame of Stella during the inbound portion of the trip has coordinate axes ct'',x''. The axes ct'',x'' are angled away from each other, which is the arrangement required for negative velocity.

For Stella, the voyage begins at A, at time $t'_A < 0$. Stella reaches the turnaround point P at time $t' = 0$. Meanwhile, Terra moves along her worldline and reaches the point B at time $t' = 0$. Between A and B, Terra's clock registers an increase $t_B - t_A$ in proper time. According to Stella, this time interval is dilated, so

$$t_B - t_A = \sqrt{1 - V^2/c^2}\,|t'_A| \tag{46}$$

After passing the turnaround point P, Stella is at rest in the t'',x''

coordinates, and she returns to Earth at the time t_D. In the time interval from $t'' = 0$ to $t'' = t''_D$, Stella observes Terra to move from C to D. In this interval, the time shown by Terra's clock increases by $t_D - t_C$ in proper time, and Stella judges this time interval to be dilated, so

$$t_D - t_C = \sqrt{1 - V^2/c^2}\, t''_D \tag{47}$$

However, the total time registered by Terra's clock includes not only the intervals AB and CD but also the interval BC. This latter interval does not correspond to negative values of t' or to positive values of t''. *The finite time interval* $t_C - t_B$ *elapses on Terra's clock in one instant* while Stella accelerates and transfers herself from one inertial frame to another at $t' = t'' = 0$.

According to Stella, in the instant between $t' = 0^-$ and $t'' = 0^+$, the hands on Terra's clock jump ahead by $t_C - t_B$. Of course, such a discontinuity is unphysical, but for this we must blame the unphysical discontinuity in the worldline of Stella: we assumed an instantaneous reversal of Stella's velocity at the point P. If we make the more reasonable assumption of a gradual change in Stella's velocity from V to $-V$, then we find that Terra's clock does not jump, but simply speeds up during the time that Stella accelerates. This speeding up of Terra's clock more than compensates for the time dilation (46) and (47) along the other parts of the worldline of Terra, and this makes Terra older when the twins are reunited, no matter what reference frame is used to calculate the aging.

3.6 THE POLE AND BARN PARADOX

The symmetric, mutual character of the length contraction also gives rise to an apparent paradox, since it seems logically contradictory that the meter rods in the spaceship reference frame are contracted relative to the meter rods in the laboratory frame and that, concurrently, the meter rods in the laboratory frame are contracted relative to those in the spaceship frame. This paradox is called the **pole and barn paradox**, because it is usually formulated by comparing the length of a barn (stationary in the laboratory frame) with the length of a long, moving pole (stationary in the spaceship frame).

Consider a barn of length 10 m in the reference frame of the laboratory, and consider a long horizontal pole carried by a fast runner (or a fast spaceship) toward this barn at high speed (see Fig. 3.17). Suppose that the pole has a length of 18 m in its own rest frame and is traveling at a speed of $V = 0.866c$. For this speed, the length contraction factor is $\sqrt{1 - (0.866)^2} = 0.50$, that is, lengths are reduced to one-half. Hence, in the reference frame of the laboratory, the length of the pole is contracted to 18 m × 0.50 = 9 m, and the pole fits inside the barn. But in the reference frame of the runner, or the spaceship, the barn is contracted to 10 m × 0.5 = 5 m, and the 18-m pole can never fit inside the barn.

The resolution of this paradox hinges on the relativity of simultaneity. At one instant of laboratory time, both the tip and the tail of the pole are inside the barn. But in the reference frame of the runner, this instant of laboratory time corresponds to two separate times: one time for the tip of the pole and a different time for the tail of the pole. According to the judgment of the runner, at the instant the tip of the pole reaches the rear wall of the barn, the tail of the pole is still outside the barn, and when the tail of the pole enters the barn, the tip has already penetrated beyond the rear wall of the barn (assuming that the rear wall is penetrable and does not stop the tip).

The Minkowski diagram shown in Fig. 3.18 helps to make this clear. The t axis is the worldline of the front of the barn, and the line AB is the worldline of the rear of the barn. The line CD is the worldline of the tip of the pole, and the line OE is the worldline of the tail of the pole. The shaded band is the region of spacetime swept out by the pole. At one instant of t time, say $t = 0$, the entire pole fits within the boundaries of the barn (see the line OP, which represents the pole at the instant $t = 0$). But at one instant of t' time, the pole does not fit within the boundaries of the barn (see the line OP', which represents the pole at the instant $t' = 0$).

An interesting complication occurs if the rear wall of the barn is completely solid, so the tip of the pole stops when it strikes the wall (Fig. 3.19). At the instant of t time that corresponds to this collision, the tail of the pole is already in the barn (see the line QR). But at the

Fig. 3.17 A runner carrying a pole runs toward a barn.

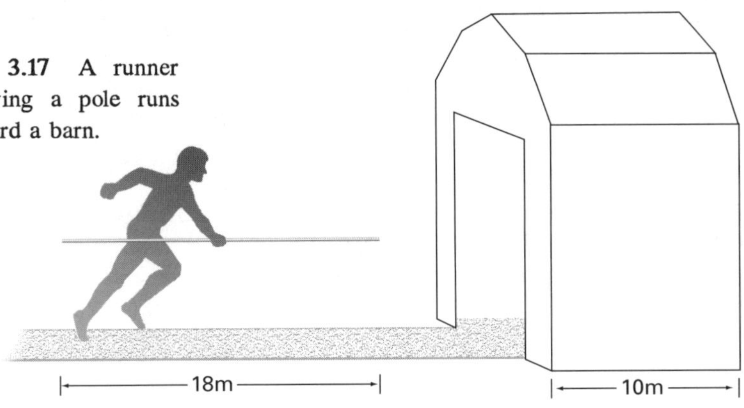

instant of t' time that corresponds to this collision, the tail of the pole is not yet in the barn (see the line QR'). If the tail of the pole were to stop instantaneously in the position R', we would have a genuine paradox, because the tail would then never enter the barn. But the tail will fail to stop there, because the collision of the tip cannot cause an immediate effect on the tail. First a compressional signal (a sound wave) must travel backward along to the pole, from tip to tail. The maximum speed of this compressional signal--and the maximum speed

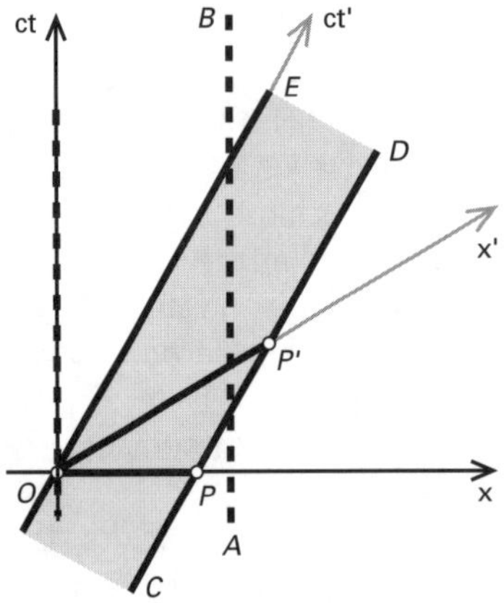

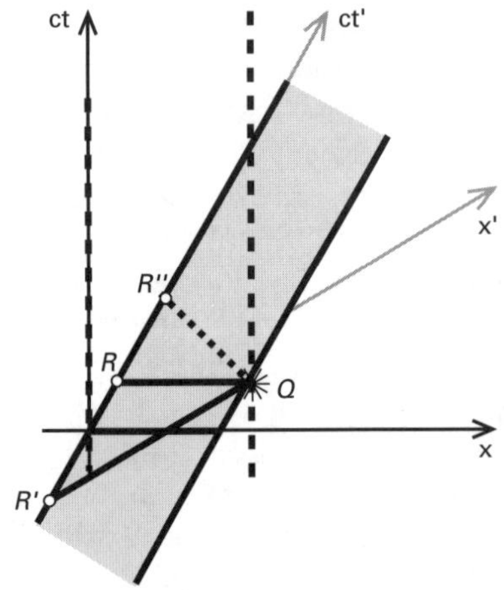

Fig. 3.18 (Left) Worldlines of the tip and the tail of the pole in motion relative to the laboratory. The body of the pole sweeps out the region of spacetime shown shaded. At the instant of time $t = 0$, the pole occupies the segment OP. At the instant $t' = 0$, the pole occupies the segment OP'.

Fig. 3.19 (Right) The tip of the pole strikes the rear wall of the barn at Q. The dotted line QR'' is the worldline of a signal traveling backward at the speed of light.

of any physical signal--is the speed of light. Figure 3.19 shows the worldline QR'' of such a signal proceeding backward from the tip of the pole at the speed of light. The worldline of this signal intercepts the worldline of the tail of the pole at the point R'', which is *inside* the barn. Thus, the tail of the pole fails to stop until it has entered the barn. The laboratory observers and the spaceship observer therefore agree on the outcome of the experiment. Both find that the tip of the pole strikes the rear of the barn and stops, and both find that the tail of the pole enters the barn and then stops. They disagree only on the time sequence of these events. For the laboratory observer, the tail end enters the barn *before* the tip strikes, whereas for the spaceship observer, the tail end enters the barn *after* the tip strikes.

REFERENCES

1. H. E. Ives and G. R. Stilwell, J. Opt. Soc. Am. **28** 215 (1938).
2. B. Rossi and D. B. Hall, Phys. Rev. **59**, 223 (1941).
3. H. Hay, J. Schiffer, T. Cranshaw, and P. Egelstaff, Phys. Rev. Lett. **4**, 165 (1960).
4. R. V. Pound and G. A. Rebka, Phys. Rev. Lett. **4**, 274 (1960).
5. J. C. Hăfele and R. E. Keating, Science **177** 168 (1962).
6. H. I. Mandelberg and L. Witten, J. Opt. Soc. Am. **52** 529 (1962).
7. D. H. Frisch and J. H. Smith, Am. J. Phys. **31**, 342 (1963).

8. W. Kündig, Phys. Rev. **129**, 2371 (1963).
9. D. S. Ayres, A. M. Cormack, A. J. Greenberg, R. W. Kenney, D. O. Cladwell, V. B. Elings, W. P. Hesse, and R. J. Morrison, Phys. Rev. D **3**, 1051 (1971).
10. J. J. Snyder and J. L. Hall, in *Laser Spectroscopy II*, ed. T. W. Hänsch et al. (Springer, Berlin, 1975).
11. J. Bailey, K. Borer, F. Combley, H. Drumm, C. Eck, F. J. M. Farley, J. H. Field, W. Flegel, P. M. Hattersley, F. Krienen, F. Lange, G. Lebee, E. McMillan, G. Petrucci, E. Picasso, O. Runolfsson, W. von Ruden, R. W. Williams, and S. Wojcicki, Nuclear Physics B **150**, 1 (1979).
12. R. F. C. Vessot, M. W. Levine, E. M. Mattison, E. L. Blomberg, T. E. Hoffman, G. U. Nystrom, B. F. Farrel, R. Decher, P. B. Eby, C. R. Baugher, J. W. Watts, D. L. Teuber, and F. D. Wills, Phys. Rev. Lett. **45**, 2081 (1980).
13. P. Juncar, C. R. Bingham, J. A. Bounds, D. J. Pegg, H. K. Carter, R. L. Mlekodaj, and J. D. Cole, Phys. Rev. Lett. **54**, 11 (1985).
14. M. Kaivola, O. Poulsen, E. Riis, and S. A. Lee, Phys. Rev. Lett. **54**, 255 (1985).
15. D. W. MacArthur, K. B. Butterfield, D. A. Clark, J. B. Donahue, P. A. M. Gram, H. C. Bryant, C. J. Harvey, W. W. Smith, and G. Comtet, Phys. Rev. Lett. **56**, 282 (1986).
16. R. Klein, R. Grieser, I. Hoog, G. Huber, I. Klaft, P. Merz, T. Kühl, S. Schröder, M. Grieser, D. Habs, W. Petrich, and D. Schwalm, Z. Phys. A **342**, 455 (1992).
17. R. W. McGowan, D. M. Giltner, S. J. Sternberg, and S. A. Lee, Phys. Rev. Lett. **70**, 251 (1993).
18. R. Grieser, R. Klein, G. Huber, S. Dickopf, I. Klaft, P. Knobloch, P. Merz, F. Albrecht, M. Grieser, D. Habs, D. Schwalm, and T. Kühl, Appl. Phys. B **59**, 127 (1994).
19. R. Grieser, T. Kühl, and G. Huber, Am. J. Phys. **63**, 665 (1995).
20. G. D. Scott and M. R. Viner, Am J. Phys. **33**, 534 (1965).
21. C. W. Sherwin, Phys. Rev. **120**, 17 (1960).
22. D. Newman, G. W. Ford, A. Rich, and E. Sweetman, Phys. Rev. Lett. **40**, 1355 (1978).
23. D. W. MacArthur, Phys. Rev. A **33**, 1 (1986).
24. U. Kraus, Am. J. Phys. **68**, 1 (2000).
25. W. Rindler, Am J. Phys. **29**, 365 (1961).

PROBLEMS

1. The orbital speed of the Earth around the Sun is 30 km/s. In 1 year, how many seconds do the clocks on the Earth lose with respect to the clocks of an inertial reference frame at rest relative to the Sun?

2. The world record for the highest speed attained by anybody relative to the Earth is 24,791 mi/h, which was attained by the crew of the *Apollo X* module on their return trip from the Moon. At this speed, what is the time dilation for a clock on the module relative to a clock on the Earth?

3. A physicist has proposed a relativistic refrigerator: you take a fresh (dead) chicken and accelerate it to a high speed so that in, say 6 months of Earth time the chicken ages by no more than 6 hours of its own time. What speed is required?

4. A deranged terrorist places a bomb equiped with a timer on a space probe just before it is launched on an interstellar mission. The timer is set to trigger the bomb after exactly 24 h. The space probe travels away from the Earth on a straight line at $v = 0.9c$. How long after launch will the observers on the Earth see the flash of light from the exploding bomb?

5. The third crew of the *Skylab* space station remained aboard for 84 days. In this time interval, how many seconds did the time dilation cause their clocks to fall behind clocks on the Earth? Take into account the fact that a clock at rest on the ground participates in the rotational motion of the Earth. Assume that the orbital speed of *Skylab* was 7.7 km/s and that the speed of a clock on the ground is the equatorial speed of rotation of the Earth, 460 m/s.

6. For the precise synchronization of the atomic clocks at the National Institute for Standards and Technology at Boulder, Colorado, and those at the Bureau International de l'Heure at Paris, France, physicists carried portable atomic clocks from one location to the other. Roughly estimate how many seconds such a traveling clock loses relative to the clocks fixed on the surface of the Earth during the trip from Boulder to Paris. The distance between Boulder and Paris is about 8000 km, and the rotational speed of the latitude of Boulder or Paris is $\simeq$300 m/s.

7. (a) In Section 3.1 we established that, in the limit $V \rightarrow 0$, the deviation between a transported clock and the synchronized clocks of the reference frame remains small even if the transport distance is large [see Eq. (14)]. The argument of Section 3.1 assumed a constant speed. Show that the result remains valid even if the transport speed is variable, provided that the path-average speed $\rightarrow 0$. (*Hint*: In the limit $V \rightarrow 0$, $\int \sqrt{1 - V^2/c^2}\, dt \simeq t - (1/2c^2) \int V\, dl$.)
(b) If you want to limit the deviation of the transported clock to less than 10^{-9} s during a clock transport over a distance of 10,000 km, what is the maximum path-average speed that you can use for the transported clock?

8. Unstable particles, such as muons, decay by a probabilistic law: the probability that a particle decays within a time t is $1 - e^{-t/\tau}$, and the probability that the particle does not decay is $e^{-t/\tau}$, where τ is the **mean lifetime** of the particle. For muons, the mean lifetime is $\tau = 2.20 \times 10^{-6}$ s. Suppose that the impact of a cosmic ray on the upper atmosphere creates a muon. The muon is initially at a height of 40 km, and it has a downward speed of $0.99c$.
(a) If there were no time dilation, what would the probability be that this muon reaches the ground at sea level?
(b) Taking time dilation into account, find the actual probability.

9. Cosmonauts want to travel from the Earth to the center of our Galaxy, at a distance of 25,000 light-years. By exploiting time dilation, the cosmonauts plan to complete this trip in 10 years of their own proper time.

(a) At what constant speed, relative to the Earth, must they travel?
(b) When they reach the center of the Galaxy, how much time has elapsed on the Earth?

10. We send a spaceship to the star Alpha Centauri, at a distance of 4.4 light-years from the Earth. Suppose this spaceship accelerates away from the Earth at 1.0 m/s^2 (as measured in the reference frame of the Earth) until it reaches the midpoint of the trip; the spaceship then decelerates at 1.0 m/s^2 until it reaches its destination. The return trip is performed in the same manner.
(a) What is the total travel time for the round trip according to the Earth clocks? Ignore the time spent at Alpha Centauri.
(b) What is the total travel time according to the clocks on the spaceship?

11. Figure 3.20 shows the worldlines of two identical clocks that move from A to B. The first clock has a straight worldline $x = vt$, where v is a constant. The second clock has the worldline $x = \frac{1}{2}at^2$, where a is a constant. For each clock, find the elapsed proper time between A and B. Which clock shows the longer proper time?

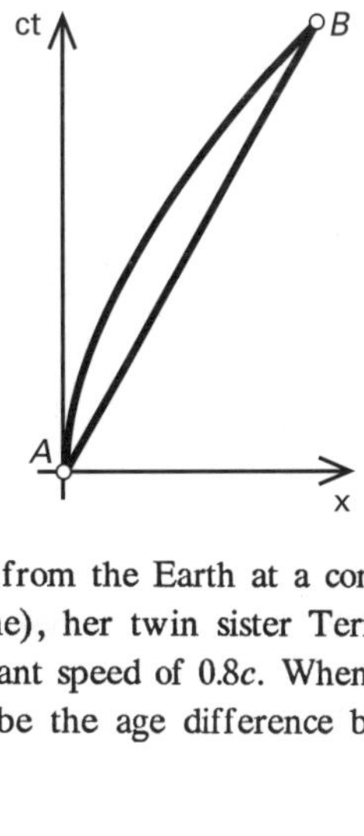

Fig. 3.20 Two worldlines connecting A and B.

12. Stella boards a spaceship that travels away from the Earth at a constant speed of $0.4c$. One year later (according to Earth time), her twin sister Terra boards a second spaceship that pursues the first at a constant speed of $0.8c$. When the second spaceship catches up with the first, what will be the age difference between the twins?

13. Consider a particle moving with constant acceleration a'_x in its own rest frame, with the equation of motion given by Eqs. (2.46) and (2.48). Show that the elapsed proper time for this particle is

$$\tau = \frac{c}{a'_x} \sinh^{-1}(a'_x t/c)$$

Verify that for short times ($a'_x t \ll c$) this coincides approximately with the ordinary time t but that for long times ($a'_x t \gg c$), it becomes $\tau \simeq c/a'_x \ln(2a'_x t/c)$, which is of the order of magnitude of c/a'_x, and hardly increases at all, regardless of how much time t elapses.

14. A spaceship launched from the Earth travels with a constant acceleration of 9.8 m/s^2 in its instantaneous rest frame (that is, a proper acceleration of 9.8 m/s^2).
(a) According to the result obtained in the preceding problem, what will be the elapsed time in the reference frame of the Earth when 1 year of *proper* time has elapsed on the spaceship? How far from the Earth will the spaceship be at that time? What will be its speed relative to the Earth at that time?
(b) Repeat the calculation for an elapsed proper time of 10 years.

15. In analogy to the light clock of Section 3.1, we can construct a bouncing-particle clock, which consists of a particle that bounces back and forth elastically between two rigid walls separated by a distance L. Suppose that the walls are at rest in the spaceship frame and that in this frame, the particle bounces up and down along the y' axis with velocity $v'_y = \pm v'$ (+ when moving upward and - when moving downward).
(a) What is the y component of the velocity of the particle in the laboratory frame?
(b) What is the time between bounces in the spaceship frame and what is the time between bounces in the laboratory frame? Are your results in agreement with the expected time dilation?

16. The light clock in Section 3.1 used a light signal bouncing between two mirrors up and down along the y' axis. Suppose that instead we use a light signal bouncing back and forth along the x' axis, between $x' = 0$ and $x' = L$. What is the time in the spaceship frame and what is the time in the laboratory frame for the forward trip? for the backward trip? for one complete round trip? Is this result in agreement with the expected time dilation?

17. The **geometrodynamic clock** of Marzke and Wheeler uses a light pulse bouncing back and forth between two mirrors, as described in Section 3.1. However, instead of being held at a fixed distance by some kind of rigid rod, the mirrors are allowed to drift freely, and their separation is checked by a trial-and error procedure involving auxiliary light signals and mirrors. In essence, the procedure hinges on the construction of parallel worldlines, since parallel worldlines in a spacetime diagram imply constant separation in space. Figure 3.21 shows the straight worldline PF of one free mirror 1 in uniform motion. To construct a worldline parallel to PF, introduce an auxiliary free mirror 2 with a worldline QF that intersects PF at some point F in the future. A light signal PQR is emitted at mirror 1, reflected by 2, and returned to 1. Introduce another auxiliary mirror 3, with worldline PG that intersects PF at P. A light signal HGF is emitted at mirror 1, reflected by 3, and returned to 1. Show that if a free particle 4 has a worldline that passes through the intersections A and B, then the worldline AB of this particle is parallel to PF. (*Hint*: This is done most easily by adopting a reference frame in which the first mirror is at rest, with a vertical worldline as drawn in Fig. 3.21.)

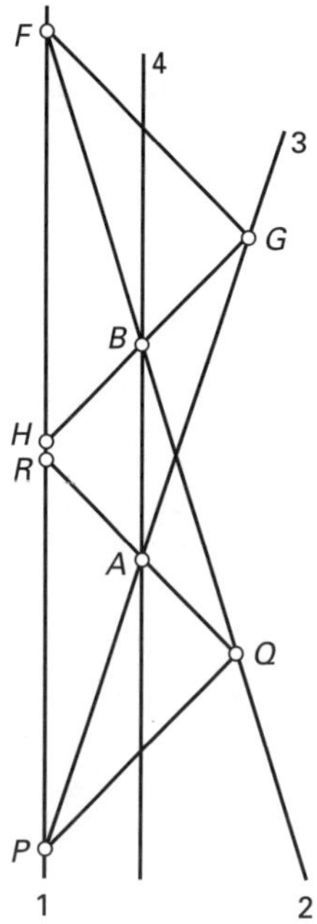

Fig. 3.21 Construction of the parallel worldlines PF and AB for the geometrodynamic clock.

18. In Section 3.2 we calculated the Doppler shift by using the reference frame of the emitter. We could equally well use the reference frame of the receiver instead. Show that this leads to

$$\nu' = \frac{1}{1 + v/c} \sqrt{1 - v^2/c^2}\ \nu$$

where the first factor is the familiar Newtonian Doppler factor for a receding emitter, and the second factor is the correction for the time dilation of the *emitter*. Show that this result coincides with Eq. (18).

19. Suppose you send a light signal of frequency ν toward a mirror that is approching you at speed v. What is the frequency of the reflected light that the mirror returns to you?

20. Very large Doppler shifts are observed in the light reaching us from distant galaxies. For instance, the light reaching us from one of the most distant galaxies has a received frequency 4.8 times smaller than the emitted frequency. What is the velocity of recession of this galaxy according to Eq. (18)? (For the large distance involved, it is actually necessary to take into account the curvature of space and time, given by General Relativity, and the naive calculation from Eq. (18) is not valid.)

21. A spaceship passes by the Sun at a speed of $0.8c$. Assume that the average wavelength of the light emitted by the Sun is 550 nm. What is the wavelength of the sunlight that the observers on the spaceship see when they are far away, approaching the Sun? when the Sun is abeam, at the point of closest approach? when they are far away, receding from the Sun?

22. As described in Section 3.3, the relativistic time-dilation effect has been tested with a two-laser technique, in which the atoms or ions in a fast beam of speed V are resonantly excited by one laser beam propagating parallel to the atomic beam and also by a second laser beam propagating in the opposite direction. Suppose that the atom or ion has two resonant frequencies ν_1 and ν_2 (in its rest frame) and that the frequencies of the light of the parallel and the antiparallel laser beams are ν_p and ν_a, respectively, in the laboratory frame.
(a) Show that if one laser is to be in resonance with ν_1 and the other laser in resonance with ν_2, then the relativistic Doppler-shift formula demands that

$$\nu_a \nu_p = \nu_1 \nu_2$$

(b) Show that the speed V of the atomic beam can be calculated from the measured frequencies,

$$V/c = (\nu_p{}^2 - \nu_1{}^2/(\nu_p{}^2 + \nu_1{}^2)$$

23. Some protons in cosmic rays have very high speeds. Consider a proton of a speed $V = 0.99c$ approaching the Earth. In its own reference frame, the proton is a sphere of radius 1×10^{-15} m. In the reference frame of the Earth, what is the longitudinal diameter of the proton? What is the transverse diameter?

24. A cannonball flies through the reference frame of your laboratory at a speed of $0.2c$. Measurement of the transverse diameter of this cannonball gives the result 0.15 m. If you perform a measurement of the longitudinal diameter of this cannonball, what will be the result?

25. What is the percent length contraction of a jet aircraft flying at twice the speed of sound, 330 m/s?

26. In the reference frame of the Sun, the speed of the Earth is 30 km/s. Assume that in its own reference frame, the Earth is a perfect sphere of radius 6.4×10^6 m. In the reference frame of the Sun, the diameter of the Earth parallel to its motion suffers a length contraction. By how many meters is the diameter contracted? By how many cubic meters is the volume contracted?

27. A cube measures 1 m $\times$ 1 m $\times$ 1 m in its own reference frame. Suppose that in the reference frame of the laboratory, this cube moves at speed $0.5c$ in a direction perpendicular to one of its faces. In the reference frame of the laboratory, what are the areas of its faces? What is its volume?

28. A meter rod is moving relative to you in a direction that makes an angle of 30^0 with its length, where the angle is measured in the reference frame of the meter rod. The speed of the meter rod is $0.5c$. What is its measured length in your reference frame?

29. Suppose we measure the equatorial circumference and the radius r of the Earth with meter rods at rest on the Earth, which (of course) participate in the rotational motion of the Earth. What difference will we find between the circumference and $2\pi r$?

30. In the reference frame of the laboratory, a right triangle of sheet metal with two 45^0 angles lies in the x-y plane, with one of its sides along the x axis (see Fig. 3.22). The length of each side is 0.20 m, and the length of the hypotenuse is $\sqrt{2} \times$ 0.20 m. Suppose that this triangle is observed from the reference frame of a spaceship moving at a speed of $0.8c$ along the x axis. What are the lengths of the sides and the hypotenuse in this reference frame? What are the angles?

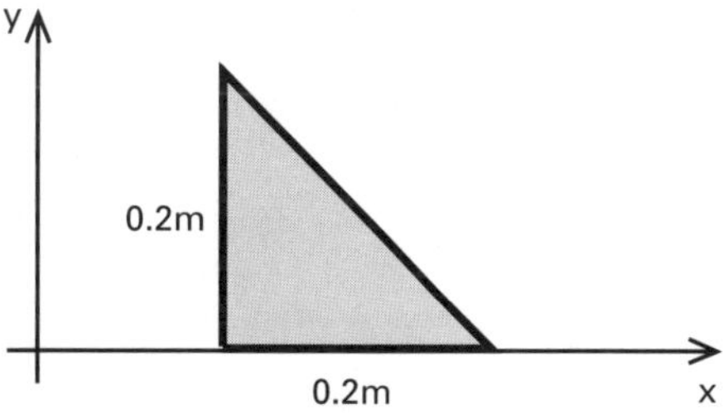

Fig. 3.22 A triangle in the x-y plane.

31. In the laboratory frame, a rectangular sheet of paper has a constant vertical velocity v_y and no horizontal velocity. The instantaneous vertical length of the paper is L (this includes the length contraction arising from the vertical velocity), and its horizontal length is $2L$. Consider the configuration of this sheet of paper at one instant of time in the spaceship frame, which has a velocity V along the x axis of the laboratory frame. Show that in this spaceship frame, the paper has the shape of a parallelogram. What are the lengths of its sides? What is the acute angle between its sides?

32. Each of two spaceships has a proper length L. One of these spaceships travels away from the Earth at a speed V. The other follows at a speed of $2V$ and passes the first. How long does the second spaceship take to pass the first? Assume that passing begins at the instant the nose of the second spaceship is even with the tail of the first and that passing is completed when the tail of the second spaceship reaches the nose of the first. Calculate the answer in (a) the reference frame of the Earth, (b) the reference frame of the first spaceship, and (c) the reference frame of the second spaceship. In which reference frame does the passing take the least time?

33. Suppose that by applying suitable external forces, we accelerate a measuring rod along the positive x axis in such a way that the rear end has a worldline

$$x = \sqrt{x_0 + c^2 t^2}$$

and a point on the rod at X meters from the rear end (as marked on the rod) has a worldline

$$x = \sqrt{(x_0 + X)^2 + c^2 t^2}$$

(a) What is the instantaneous speed $v = dx/dt$ of the point X?
(b) What is the acceleration a'_x of the point X in its instantaneous rest frame? What point X on the measuring rod has the highest acceleration?
(c) Evaluate dx/dX at a fixed value of t, and use part (a) to show that the result can be written

$$\frac{dx}{dX} = \sqrt{1 - v^2/c^2}$$

This establishes that each segment dX of the rod is contracted by the usual length-contraction factor when observed in the laboratory frame; that is, each segment dX of the rod is contracted in the way expected for an isolated segment, accelerating on its own without any compressional effect from the adjacent segments. Accordingly, a measuring rod accelerated in this way has no internal stresses–it has its "natural" length.
(d) If the rod extends from $X = 0$ to $X = L$, what is its instantaneous length in the laboratory frame at time t?

34. Try to use the Loedel diagram (see Problem 2.16) for a graphical explanation of the twin paradox, as in Fig. 3.16. Can the Loedel diagram be used for this purpose? Why not?

35. Two flywheels are mounted on a rigid base, and a flexible drive belt runs over these flywheels (see Fig. 3.23). In the reference frame of the base, when the flywheels are rotating, the top portion of the drive belt travels toward the right at speed v, and the bottom portion travels toward the left at the same speed v. Thus, the top and bottom portions of the belt experience length contraction, which tightens the belt around the flywheels. However, in a reference frame moving toward the right with speed v, the base experiences length contraction, which ought to loosen the belt around the flywheels. Resolve this paradox by a qualitative argument.* (*Hint*: Consider the lower portion of the belt as seen in the reference frame moving with the upper portion.)

---

* I am indebted to J. Traschen for this problem.

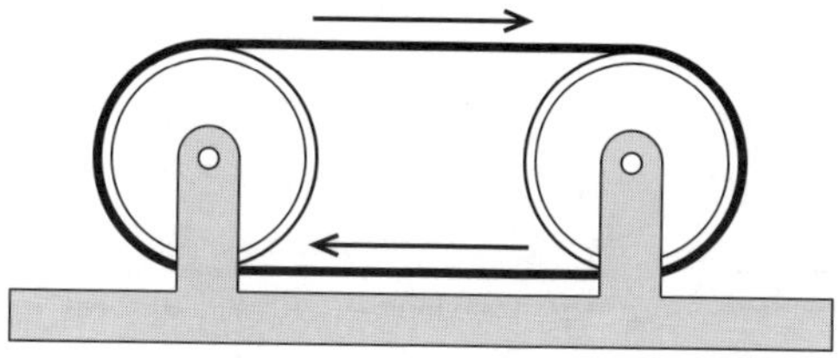

Fig. 3.23 Two flywheels connected by a drive belt.

36. In the laboratory frame, a meter rod has a large constant horizontal velocity v_x and also a small constant downward vertical velocity v_y. The meter rod approaches an open manhole in the street, which has a diameter of only 0.6 m (see Fig. 3.24a). If v_x is large, the length contraction makes the meter rod shorter than 0.6 m, and it can enter the manhole. But in the spaceship frame traveling along the x axis with speed $V = v_x$, the manhole suffers length contraction, and it would seem that the meter rod cannot enter the manhole. Resolve this paradox by a qualitative argument, showing that in the reference frame of the spaceship, the meter rod is tilted down at its front end and therefore passes through the manhole at an angle (see Fig. 3.24b).

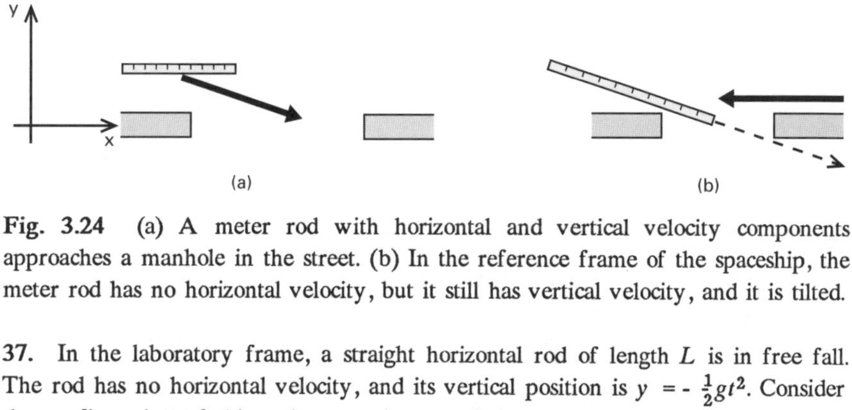

Fig. 3.24 (a) A meter rod with horizontal and vertical velocity components approaches a manhole in the street. (b) In the reference frame of the spaceship, the meter rod has no horizontal velocity, but it still has vertical velocity, and it is tilted.

37. In the laboratory frame, a straight horizontal rod of length L is in free fall. The rod has no horizontal velocity, and its vertical position is $y = -\frac{1}{2}gt^2$. Consider the configuration of this rod at one instant of time in the spaceship frame which has a velocity V along the x axis of the laboratory frame. What is the equation $y' = y'(x', t')$ that describes the instantaneous position of points on the rod in the spaceship frame? Plot y' as a function of x' for $t' = 0$. Is the rod straight? Does the shape of the rod remain fixed as it continues to fall? Does the length of the rod remain fixed?

38. A straight rod of proper length L slides at high speed in a close-fitting, long, horizontal tube that has a gap of length L at its bottom. In the reference frame of the tube (or the laboratory frame), the rod is length-contracted, it is shorter than the gap, it falls out of the tube, and it then accelerates downward with the free-fall acceleration g (see Fig. 3.25a). But in the reference frame of the rod (or the spaceship frame), the tube and the gap are length-contracted, and the rod is too long to fall out. Resolve this apparent paradox.* (*Hint*: At one instant of spaceship

time, the configuration of the rod is as shown in Fig. 3.25b. Explain this bending of the rod in the spaceship frame.)

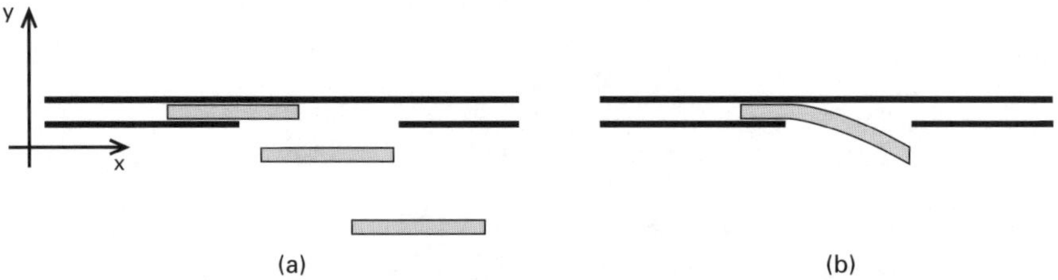

Fig. 3.25 (a) In the reference frame of the tube, the speeding rod is length-contracted and it falls out of the hole. The figure shows three successive snapshots of the falling rod. (b) In the reference frame of the rod, the tube and the hole are length-contracted. The rear end of the rod is still in the tube, but the front end has already fallen some vertical distance.

* This paradox is due to Rindler [25]; I thank an anonymous correspondent for bringing it to my attention.

4

Relativistic Mechanics

IF I HAD ONLY KNOWN, I SHOULD HAVE BECOME A WATCHMAKER.
Albert Einstein, commenting on the role of relativity in the making of the atom bomb.

The Newtonian laws of mechanics and the definitions of momentum and energy are intimately related to the Newtonian concepts of space and time. Because Special Relativity imposes drastic modifications of the concepts of space and time, it also imposes corresponding modifications of mechanics. In this chapter we will formulate the relativistic laws of mechanics and the relativistic definitions of momentum and energy. In this formulation of relativistic mechanics, we will endeavor to retain as much of Newtonian mechanics as possible, and we will reject a Newtonian law or formula only if we are compelled to do so ("If it ain't broke, don't fix it").

4.1 THE LAWS OF MOTION

Newtonian mechanics rests on Newton's three laws of motion: (*i*) In the absence of external forces, a body at rest remains at rest, and a body in motion continues to move at uniform velocity; (*ii*) An external force acting on a body causes an acceleration proportional to the force, $m\mathbf{a} = \mathbf{F}$; (*iii*) For every force there is a reaction force of equal magnitude and opposite direction. How do we have to reformulate these Newtonian laws to bring them into accord with relativity?

The First Law is equally valid in Newtonian mechanics and relativistic mechanics. Inertial reference frames are defined in the same way in Newtonian and relativistic mechanics, and the conditions of rest and of uniform motion have the same meanings. If a body is in a state of uniform motion in one inertial reference frame, then it will be in a state of uniform motion in any other inertial reference frame, and this is true regardless of whether the transformation from one reference frame to another is a Galilean transformation or a Lorentz transformation. Thus, Newton's First Law is already consistent with relativity, and it remains invariant when we transform from one inertial reference frame to another. This law requires no modification.

However, the Second Law is inconsistent with relativity. We know that in an inertial reference frame, Newtonian mechanics is in excellent agreement with observation and experiment in the limiting case of low particle speeds, which means that $m\mathbf{a} = \mathbf{F}$ is valid for low speeds. However, if $m\mathbf{a} = \mathbf{F}$ were to be valid also for arbitrarily large speeds, then a constant force acting on a particle would cause a constant acceleration, and the speed of the particle would increase indefinitely, ultimately exceeding the speed of light. As we saw in Chapter 2, particle speeds in excess of the speed of light are not acceptable in relativity, since they lead to intolerable conflicts with causality. To avoid these conflicts, we need a new equation of motion such that a constant force produces an acceleration that gradually decreases as the speed increases (as in the example of hyperbolic motion in Section 2.6), so as to prevent the particle from ever attaining the speed of light. If we reject $m\mathbf{a} = \mathbf{F}$, what replaces it? An obvious candidate is Newton's Second Law formulated as an equation for the rate of change of momentum, $d\mathbf{p}/dt = \mathbf{F}$. In Newtonian mechanics, $\mathbf{p} = m\mathbf{v}$, so $d\mathbf{p}/dt$ is equal to $m\mathbf{a}$, and the difference between the two formulations of Newton's Second Law is merely a question of notation. But, as we will see in Section 4.2, in relativistic mechanics we have to adopt a modified definition of momentum, and then $d\mathbf{p}/dt$ is not equal to $m\mathbf{a}$.

The Third Law is also inconsistent with relativity. The action and reaction forces act on different particles, which might be separated by some distance. Such a separation of the points of application of the action and reaction forces leads to inconsistencies for the simultaneous balance of these forces. The relativity of simultaneity implies that if action and reaction are balanced in one reference frame at one instant of time, then in another reference frame, action and reaction are balanced at *different* instants of time, and they are not necessarily balanced at the same instant of time. But all is not lost. Although we must reject the Third Law for forces that act between particles separated by a distance, we can retain the Third Law for contact forces, where action and reaction act at essentially the same place. Thus, we can retain the Third Law for the contact forces that act during collisions between pointlike particles.

We can then summarize the status of the three laws of motion in relativity as follows: the First Law remains valid, the Second Law becomes $d\mathbf{p}/dt = \mathbf{F}$ with a modified definition of momentum, and the Third Law is restricted to contact forces.

4.2 RELATIVISTIC MOMENTUM

To discover the relativistic formula for momentum, we begin with the assumption that the direction of the momentum of a particle is parallel to its velocity **v**, as in Newtonian mechanics. This assumption is self-evident, since the direction of the velocity is the only preferred direction associated with the motion of a free particle. However, in contrast to Newtonian mechanics, we assume that the magnitude of the momentum is not mv but some more complicated function of the magnitude of the velocity. Thus, we write

$$\mathbf{p} = f(v^2)m\mathbf{v} \tag{1}$$

where $f(v^2)$ is some function of the magnitude of the velocity. The function is subject to the restriction $f(v^2) \to 1$ for $v^2 \to 0$; this restriction ensures that for low, nonrelativistic velocities, formula (1) coincides with the Newtonian formula.

We can determine the function $f(v^2)$ by analyzing an elastic collision between two pointlike particles. Since the Third Law is valid for such a collision, the momentum changes of the two particles are equal and opposite, and the net momentum is conserved. We will see that if this conservation of momentum is to be valid in all inertial reference frames, in accordance with the Principle of Relativity, then the function $f(v^2)$ in Eq. (1) must have a very special form.

Figure 4.1 shows the collision in the laboratory frame. The momenta of the two particles 1 and 2 have been carefully selected to give a highly symmetric collision, and it is this symmetry that will permit us to draw conclusions about the function $f(v^2)$. Particle 1 has both x and y components of momentum, but particle 2 has only a y component. Before the collision (Fig. 4.1a), particle 1 has a negative y component of momentum, and particle 2 has a positive y component of the same magnitude. Momentum conservation requires that after the collision, particle 1 have a positive y component and particle 2 a negative y component. Thus, the particles merely reverse (or inter-

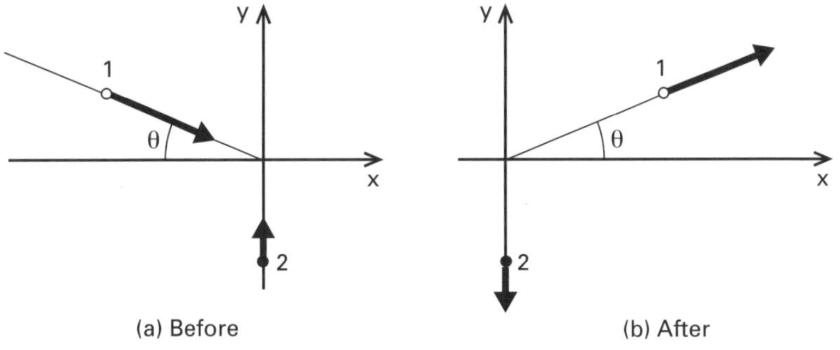

Fig. 4.1 Collision between particles 1 and 2 as seen in the laboratory frame. (a) Before the collision. (b) After the collision.

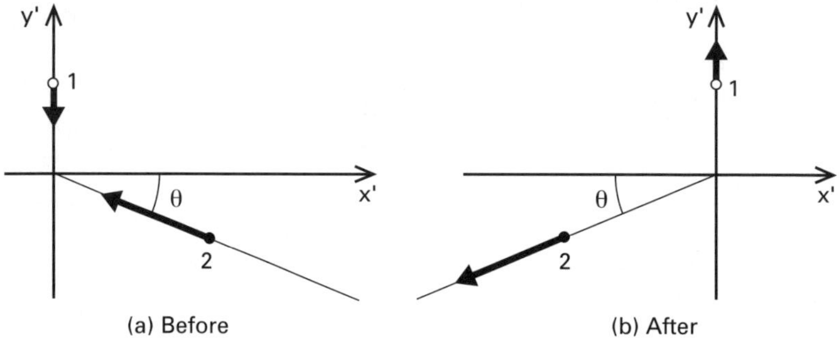

Fig. 4.2 Collision between particles 1 and 2 as seen in the spaceship frame. (a) Before the collision. (b) After the collision.

change) their y components of momentum during the collision, but they keep their x components of momentum unchanged.

Figure 4.2 shows the collision in the spaceship frame, which moves toward the right with a velocity that matches the x component of the velocity of particle 1. In this spaceship frame, particle 1 has only a y' component of momentum, but particle 2 has both x' and y' components. As seen in the spaceship frame, momentum conservation requires that the particles again reverse their y' components of momentum, but keep their x' components unchanged. Note the symmetry between the two reference frames: Except for a contrary direction, whatever particle 1 does in the laboratory frame, particle 2 does in the spaceship frame, and vice versa.

It is convenient to express momentum conservation in terms of the changes of momentum of the particles. In the laboratory frame, the changes in the y components of the momenta of the particles are $\Delta p_{1,y}$ and $\Delta p_{2,y}$, and conservation of momentum requires that these changes be opposite:

$$\Delta p_{1,y} = -\Delta p_{2,y} \tag{2}$$

In view of the symmetry between the behavior of the particles in the two reference frames, the momentum change of particle 2 in the laboratory frame matches the momentum change of particle 1 in the spaceship frame, with a reversed sign:

$$\Delta p_{2,y} = -\Delta p'_{1,y} \tag{3}$$

Substituting this into Eq. (2), we find

$$\Delta p_{1,y} = \Delta p'_{1,y} \tag{4}$$

The change in the y or y' component of the momentum equals -2 times the initial component of momentum. By means of Eq. (1), we can then write Eq. (4) as

$$-2f(v^2)mv_y = -2f(v'^2)mv'_y \tag{5}$$

Since particle 1 has no x component of velocity in the spaceship frame, $v'^2 = v'_y{}^2$, and we can rewrite Eq. (5) as

$$2f(v^2)mv_y = 2f(v'_y{}^2)mv'_y \tag{6}$$

The relativistic velocity combination formula for the y component of the velocity, Eq. (2.38), tells us that, with $V = v_x$,

$$v'_y = v_y \frac{\sqrt{1 - V^2/c^2}}{1 - Vv_x/c^2} = \frac{v_y}{\sqrt{1 - v_x{}^2/c^2}} \tag{7}$$

Substituting this into Eq. (6), we find

$$2f(v^2)mv_y = 2f\left(\frac{v_y{}^2}{1 - v_x{}^2/c^2}\right) \frac{mv_y}{\sqrt{1 - v_x{}^2/c^2}}$$

If we now proceed to the limit $v_y \rightarrow 0$ (while keeping v_x large), we obtain

$$2f(v^2)mv_y = 2f(0) \frac{mv_y}{\sqrt{1 - v^2/c^2}}$$

Taking into account that $f(0) = 1$, we conclude from this equation that $f(v^2) = 1/\sqrt{1 - v^2/c^2}$. With this result for the function $f(v^2)$, Eq. (1) becomes

$$\boxed{\mathbf{p} = \frac{m\mathbf{v}}{\sqrt{1 - v^2/c^2}}} \tag{8}$$

With the notation $\gamma = 1/\sqrt{1 - v^2/c^2}$ introduced in Chapter 2, we can also write this as

$$\mathbf{p} = m\gamma\mathbf{v} \tag{9}$$

Thus, by exploiting the Principle of Relativity and the relativistic combination formula for velocity, we have shown that what is conserved in this collision is not the Newtonian momentum but a modified momentum. Equation (8) gives us this modified, **relativistic momentum.**

Figure 4.3 is a plot of the magnitude of the relativistic momentum as a function of the speed. For low speeds, the momentum increases in proportion to v, as in the Newtonian case. But for v close to c, the relativistic momentum is much larger than the Newtonian momentum,

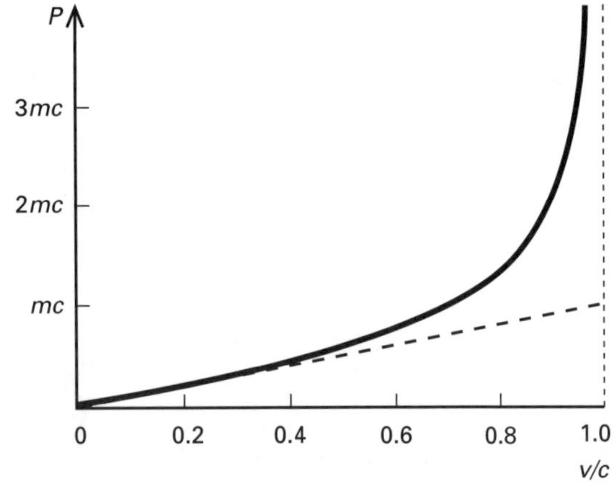

Fig. 4.3 Relativistic momentum vs. speed. For comparison, the dashed line gives the nonrelativistic momentum mv.

and for $v = c$, the momentum becomes infinite. This infinite value of the momentum ensures that when a finite force acts on a particle for a finite time, the final speed remains always less than c.

The relativistic momentum is sometimes written as

$$\mathbf{p} = m(v)\mathbf{v} \tag{10}$$

where $m(v)$ is a "velocity-dependent mass" defined as

$$m(v) = \frac{m}{\sqrt{1 - v^2/c^2}} \tag{11}$$

In contrast, the ordinary mass m that appears on the right side is called the **rest mass**. Evidently, the rest mass equals the velocity-dependent mass at zero velocity, $m = m(0)$, and to emphasize that the rest mass m is a constant, it is sometimes written as m_0. But introducing a velocity-dependent mass into the equations of relativistic mechanics does more harm than good, and we will avoid it.* In all the formulas of this book, the mass m is always the rest mass.

With our formula for the relativistic momentum, we can write Newton's Second Law as

* Instead of associating the factor $\sqrt{1 - v^2/c^2}$ with the mass, it is better to write $\mathbf{v}$ in Eq. (8) as a derivative, $\mathbf{v} = d\mathbf{x}/dt$, and to associate this factor with the differential dt that then appears in the denominator of Eq. (8). The product then becomes the proper time interval, $\sqrt{1 - v^2/c^2}\,dt = d\tau$, and $\mathbf{p} = m d\mathbf{x}/d\tau$. See the definition of the energy-momentum four-vector in Section 4.5.

$$\frac{d\mathbf{p}}{dt} = \mathbf{F} \tag{12}$$

or

$$\boxed{m\frac{d}{dt}\frac{\mathbf{v}}{\sqrt{1 - v^2/c^2}} = \mathbf{F}} \tag{13}$$

This is the relativistic equation of motion for a particle subjected to a force **F**. For low speeds, the square root in Eq. (13) can be neglected, and then the relativistic equation of motion reduces to the familiar form of Newton's Second Law, mdv/dt = **F**. But for high speeds, the relativistic equation of motion differs drastically from Newton's Law. We will use the relativistic equation of motion in Section 4.6 to obtain the motion of a high-speed particle in a constant electric field and in a constant magnetic field.

For all the moving bodies that we encounter in everyday life and all the bodies that engineers normally deal with--basketballs, baseballs, automobiles, ships, TGV trains, aircraft, rockets, Space Shuttles, artificial satellites, etc.--the speeds are very low compared with the speed of light, and relativistic effects are insignificant. The motion of all such bodies can be adequately described by the laws of Newtonian, nonrelativistic mechanics. However, as the following example illustrates, engineers sometimes have to consider relativistic effects in devices that contain high-speed particles, such as high-energy electrons.

■ **EXAMPLE 1.** The electrons in the beam of a modern, high-voltage (15-kV) TV tube have a speed of $0.24c$. For this speed, what is the difference between the magnitudes of the relativistic and the nonrelativistic Newtonian momentum?

SOLUTION. The difference between the relativistic and the Newtonian expressions is

$$\frac{mv}{\sqrt{1 - v^2/c^2}} - mv$$

and the fractional difference is

$$\frac{1}{\sqrt{1 - v^2/c^2}} - 1 = \frac{1}{\sqrt{1 - (0.24)^2}} - 1 = 1.030 - 1 = 0.030$$

Thus, the difference is about 3%.

The preceding example indicates that electronics engineers must take relativity into account in the design of high-voltage TV tubes and X-ray tubes and also in the design of high-voltage electron microscopes and high-power generators of radio waves (klystrons). Of course, relativity is also essential in the design of particle accelerators, in which particles attain speeds very close to the speed of light. Most accelerators are used for pure research, but some are used for practical applications, such as the production of proton and neutron beams for cancer therapy and the production of radioisotopes for medical and industrial use. For instance, after construction of the SCC accelerator near Dallas, Texas, was canceled, the linear accelerator that had already been assembled there, and that had been intended to serve as the first stage of the SCC, was acquired by a commercial organization and adapted for the production of rare radioisotopes.

4.3 RELATIVISTIC KINETIC ENERGY

Corresponding to our new definition of the relativistic momentum, we will need a new definition of kinetic energy. In Newtonian mechanics, the kinetic energy is defined by the requirement that the change of kinetic energy equal the work done by the external force acting on the particle (the work-energy theorem). In relativistic mechanics, we impose the same requirement. Thus, the change of the relativistic kinetic energy must equal the work done by the force. Accordingly, for a small displacement $d\mathbf{x}$, the change of relativistic kinetic energy is

$$dK = \mathbf{F} \cdot d\mathbf{x} = \frac{d\mathbf{p}}{dt} \cdot d\mathbf{x} = d\mathbf{p} \cdot \mathbf{v} \tag{14}$$

From the relativistic formula for momentum, we find

$$d\mathbf{p} = m\, d\left(\frac{\mathbf{v}}{\sqrt{1 - v^2/c^2}}\right) = m\,\frac{d\mathbf{v}}{\sqrt{1 - v^2/c^2}} + \frac{\frac{1}{2} m\mathbf{v}\, d(v^2)/c^2}{(1 - v^2/c^2)^{3/2}} \tag{15}$$

so

$$dK = \mathbf{v} \cdot d\mathbf{p} = \frac{\frac{1}{2} m\, d(v^2)}{\sqrt{1 - v^2/c^2}} + \frac{\frac{1}{2} m v^2 d(v^2)/c^2}{(1 - v^2/c^2)^{3/2}}$$

Combining the two terms on the right side with a common denominator, we obtain

$$dK = \frac{\frac{1}{2} m\, d(v^2)}{(1 - v^2/c^2)^{3/2}}$$

or

$$dK = d\left(\frac{mc^2}{\sqrt{1 - v^2/c^2}}\right) \tag{16}$$

The expression on the right side is a complete differential, which makes it easy to integrate dK. When we integrate from a lower limit $K = 0$ and $v^2 = 0$ to an upper limit $K \neq 0$ and $v^2 \neq 0$, we obtain

$$K = \int_0^{v^2} d\left(\frac{mc^2}{\sqrt{1 - v^2/c^2}}\right) = \left[\frac{mc^2}{\sqrt{1 - v^2/c^2}}\right]_0^{v^2} \tag{17}$$

Thus,

$$\boxed{K = mc^2\left(\frac{1}{\sqrt{1 - v^2/c^2}} - 1\right)} \tag{18}$$

or, with the notation $\gamma = 1/\sqrt{1 - v^2/c^2}$,

$$K = mc^2(\gamma - 1) \tag{19}$$

This is the formula for the **relativistic kinetic energy.**

To compare this with the Newtonian formula, we use the familiar approximation

$$\frac{1}{\sqrt{1 - v^2/c^2}} \simeq 1 + \frac{1}{2}\frac{v^2}{c^2} + \ldots \tag{20}$$

which is valid if $v/c << 1$. With this approximation, Eq. (18) reduces to

$$K \simeq mc^2\left(1 + \frac{v^2}{2c^2} - 1\right) \simeq \frac{1}{2}mv^2 \tag{21}$$

Thus, for low velocities, the relativistic and Newtonian formulas coincide.

Figure 4.4 is a plot of the relativistic kinetic energy as a function of speed. For low speeds, the relativistic kinetic energy increases in proportion to v^2, as in the Newtonian case. But for high speeds, the relativistic kinetic energy becomes much larger than the Newtonian value. For $v = c$, the relativistic kinetic energy, like the relativistic momentum, becomes infinite. This, again, tells us that the speed of light is unattainable for a particle, since we cannot supply an infinite amount of energy to the particle.

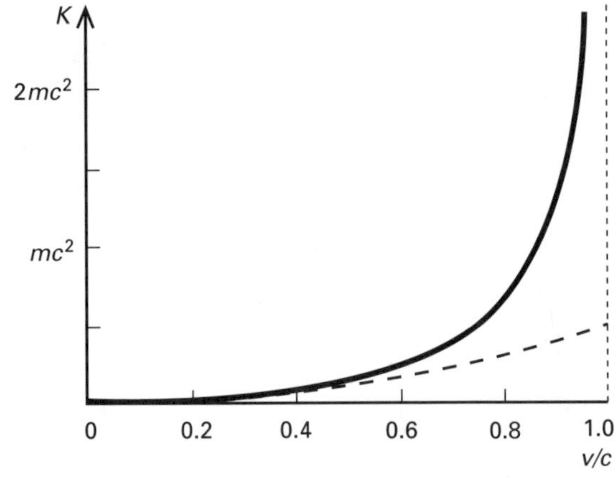

Fig. 4.4 Relativistic kinetic energy vs. speed. For comparison, the dashed line gives the nonrelativistic kinetic energy $\frac{1}{2}mv^2$.

The kinetic energies of the high-energy particles produced by accelerators are usually measured in MeV, GeV, or TeV, that is, in millions eV, billions eV, or trillions eV (see Table 4.1). For instance, the Stanford Linear Accelerator (SLAC; see Fig. 4.5) and the CERN LEP accelerator (see Fig. 4.6) produce electrons and antielectrons of 50 MeV, and the Fermilab accelerator (see Figs. 4.7 and 5.4) produces protons and antiprotons of 1 TeV.

Fig. 4.5 The Stanford Linear Accelerator. Electrons are accelerated in a 3-km tunnel that extends from the upper right toward the lower left. (SLAC)

Fig. 4.6 The underground tunnels of the CERN accelerators are marked by the circles in this photograph of an area near Geneva. The smallest circle shows the location of the PS accelerator, the middle circle shows the SPS accelerator, and the large circle shows the LEP accelerator, 27 km in circumference. (CERN photo)

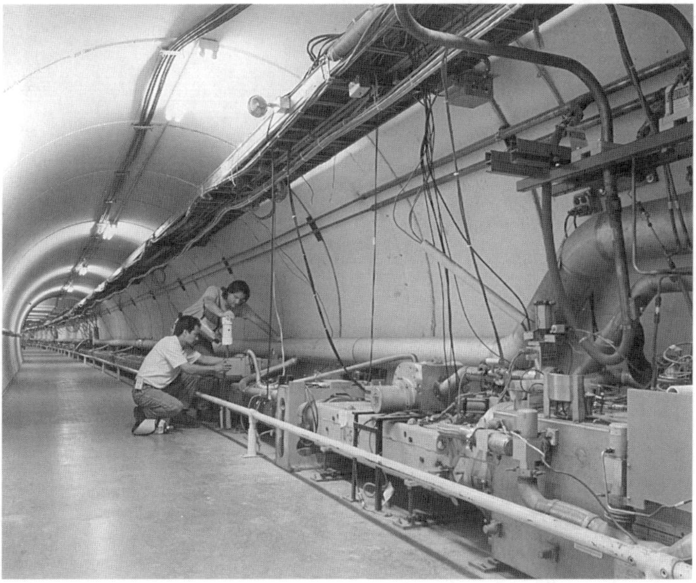

Fig. 4.7 Interior of the tunnel of the Fermilab accelerator, a circle of circumference 6.3 km. The electromagnets (at bottom) produce a magnetic field which holds the protons in an evacuated beam pipe placed inside them. (Fermilab)

An electron of a kinetic energy of 1 MeV (or a proton of kinetic energy 2 GeV) has a speed of about $0.9c$, and an electron of a kinetic energy of 10 or more MeV (or a proton of 20 or more GeV) has a speed very nearly equal to the speed of light.

TABLE 4.1 ENERGY UNITS FOR HIGH-ENERGY PARTICLES

1 MeV = 10^6 eV = 1.6×10^{-13} J
1 GeV = 10^9 eV = 1.6×10^{-10} J
1 TeV = 10^{12} eV = 1.6×10^{-7} J

EXAMPLE 2. The maximum kinetic energy that electrons attain at the Stanford Linear Accelerator (SLAC) is 50 GeV. How close to the speed of light is the speed of an electron of this energy?

SOLUTION. We begin by solving Eq. (18) for $\sqrt{1 - v^2/c^2}$,

$$\sqrt{1 - v^2/c^2} = \frac{mc^2}{K + mc^2} \tag{22}$$

Since the speed is very near c, we can make the approximation

$$\sqrt{1 - v^2/c^2} = \sqrt{1 + v/c}\,\sqrt{1 - v/c} \simeq \sqrt{2}\,\sqrt{1 - v/c} \tag{23}$$

Squaring Eq. (22), we then find

$$1 - \frac{v}{c} \simeq \frac{1}{2}\left(\frac{mc^2}{K + mc^2}\right)^2 \tag{24}$$

The quantity mc^2 has the dimensions of energy, and it is convenient to express it in units of eV,

$$\begin{aligned} mc^2 &= (3.00 \times 10^8 \text{ m/s})^2 \times 9.11 \times 10^{-31} \text{ kg} = 8.20 \times 10^{-14} \text{ J} \\ &= 8.20 \times 10^{-14} \text{ J} \times (1 \text{ eV}/1.60 \times 10^{-19} \text{ J}) = 5.11 \times 10^5 \text{ eV} \end{aligned}$$

so

$$1 - \frac{v}{c} \simeq \frac{1}{2}\left(\frac{5.11 \times 10^5 \text{ eV}}{50 \times 10^9 \text{ eV} + 5.11 \times 10^5 \text{ eV}}\right)^2 = 5.2 \times 10^{-11}$$

Thus, the speed is $0.99999999948c$.

Tests of the relativistic formula for the momentum, or the variation of "mass" with speed, were first performed by Kaufmann in 1901, who experimented with high-speed electrons, or beta rays, emitted by radioactive sources. He determined the momentum of the electrons from the radius of their circular orbit in a uniform magnetic field, and he determined their speed by means of a "velocity filter" consisting of perpendicular magnetic and electric fields (see Section 4.6 for a discussion of the orbital motion of relativistic particles in a magnetic field). These and other early experiments by Bucherer, Guye and Lavanchy, and others indicated an increase of mass with speed, in apparent agreement with the relativistic formula (11). But according to a reanalysis of the experimental procedures, the measurement errors in most of these experiments were much larger than was originally claimed, and none of the early experiments provided a reliable quantitative confirmation of the relativistic formula [1, 2].

With modern laboratory equipment, it is fairly easy to confirm the relativistic formulas for momentum and energy to within a few percent [3, 4], but it is difficult to achieve higher precision because of the uncertainties in the measurements of the momentum and speed (or energy) of individual particles. The most precise experiments were performed with high-energy protons produced by accelerators, and these confirmed the relativistic formula to within 0.1% [5, 6]. As in the earlier experiments, the momentum of the protons was determined from the observed radius of their circular orbit in a magnetic field, but the speed was determined directly from the period of the orbit.

Although the relativistic formulas for energy and momentum are not amenable to direct tests of high precision, there are some aspects of these formulas that have been tested with very high precision. For instance, one consequence of Eq. (18) is that the energy becomes infinite as the speed approaches the speed of light, stated otherwise, the speed becomes independent of the energy as the energy becomes large. This prediction of relativity was tested by comparing the travel times of high-energy electrons and of short pulses of visible light; the speeds were found to be equal to within the experimental error, about 3 parts in 10^6 [7]. Even higher precision was attained in an elegant experiment at the Stanford Linear Accelerator [8], in which electrons of energies between 15 and 20 GeV were made to race against photons generated by the penetration of the electrons through a thin target of metal (*Bremsstrahlung*). The electrons and the photons emerging from the target raced a distance of 1000 m, and it was found that at the finish, the photons were not perceptibly ahead of the electrons. This experiment established that electrons of 15 to 20 GeV have nearly equal speeds, and nearly equal to the speed of light, to within 2 parts in 10^7.

An even more precise indirect test of the relativistic formulas for energy and momentum has emerged from measurements of the intrinsic magnetic moment of the electron (measurement of the g factor of the electron) [9]. These high-precision measurements determined the magnetic moment by a comparison of the frequency of precession of the magnetic moment in a magnetic field with the frequency of the simultaneous orbital motion of the electron in this same magnetic field. One such measurement was performed with electrons of very low energy (5×10^{-4} eV) [10]; another was performed with electrons of relativistic energy (0.11 MeV) [11]. The comparison between the magnetic moments obtained in these two measurements established that the quantity $\sqrt{1 - v^2/c^2}(p/m)(dp/dK) - 1$, which enters into the comparison, is zero to within 5 parts in 10^9. According to the relativistic formulas for energy and momentum, this quantity is *exactly* zero. Thus, these measurements provide us with an extremely precise confirmation of the value of a quantity constructed from the relativistic energy and momentum, which implies an indirect confirmation of the relativistic formulas. Similar measurements have been performed on muons, with somewhat lower precision.

Table 4.2 summarizes experimental tests of relativistic dynamics. This table also includes some tests that apply only to particles of zero mass (photons and neutrinos), to be discussed in Section 4.5.

TABLE 4.2 TESTS OF RELATIVISTIC DYNAMICS

Experimenters	Year	Fractional Experimental Uncertainty
Kaufmann and others (p vs. v, electrons)	1901-1921	large
Grove and Fox (p vs. v, protons) [5]	1953	10^{-3}
Zrelov et al. (p vs. v, protons) [6]	1958	10^{-3}
Brown et al. ($v \to c$, electrons) [7]	1973	3×10^{-6}
Guiragossian et al. ($v \to c$, electrons) [8]	1975	2×10^{-7}
Newman et al. (g factor, electrons) [9]	1978	5×10^{-9}
Longo (v vs. E, neutrinos) [12]	1987	2×10^{-9}
Haines et al. (v vs. E, neutrinos or photons) [13]	1990	2×10^{-13}
Warner and Nather (v vs. E, photons) [14]	1969	5×10^{-17}
Schaefer (v vs. E, photons) [15]	1999	6×10^{-21}

4.4 THE EQUIVALENCE OF ENERGY AND MASS

The equivalence of energy and mass, according to the formula $E = mc^2$, is probably the most famous of Einstein's contributions to physics. Einstein presented several arguments leading to this equivalence. His early arguments were based on examination of the energy and momentum changes during *a* process of emission of light by some

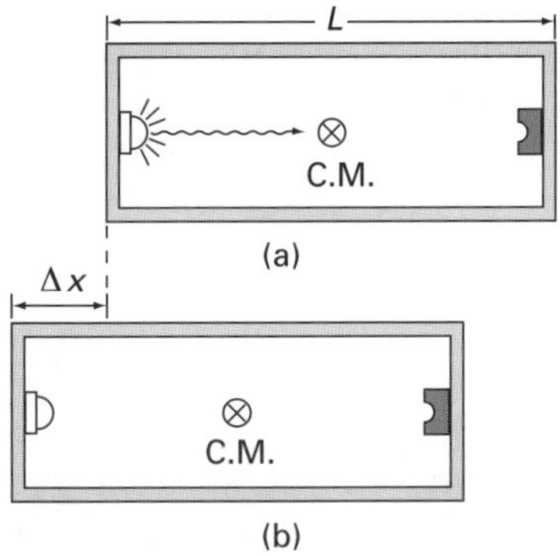

Fig. 4.8 (a) A closed box, with an emitter of light on the left and an absorber on the right. (b) The recoil of the box gives it a displacement Δx toward the left.

source. One of these arguments relies on a simple *Gedankenexperiment** involving an emitter and an absorber of light enclosed in a box of mass M and length L. The emitter is attached to one end of the box, the absorber to the opposite end (see Fig. 4.8a). From Maxwell's theory it is known that a plane light wave of energy ΔE carries a momentum $p = \Delta E/c$ (this momentum of the light wave is closely related to the "pressure of light"; the pressure is simply a momentum transfer per unit area). Imagine that the emitter emits a pulse of light of energy ΔE toward the right. The emitter and the box to which it is attached then suffer a recoil toward the left. If the box is initially at rest, it acquires a recoil velocity toward the left,

$$v_x = -\frac{p}{M} = -\frac{\Delta E}{Mc} \tag{25}$$

The light pulse takes a time of about $t = L/c$ to travel from one end of the box to the other. When the light pulse is absorbed, the absorber and the box to which it is attached suffer a recoil toward the right, which stops the motion of the box and leaves it in a displaced position (see Fig. 4.8b). The displacement of the box toward the left is

$$\Delta x = v_x t = -\frac{\Delta E}{Mc} \times \frac{L}{c} \tag{26}$$

But in an isolated system, the center of mass must remain stationary. The transfer of energy from the left end of the box to the right must therefore be accompanied by the displacement of some amount of Δm mass from the left to the right, so the center of mass of the system remains stationary,

---

* German for "thought experiment," that is, an imagined, idealized experiment whose outcome can be predicted by thought alone.

$$M\Delta x + (\Delta m)L = 0$$

or

$$-L\Delta E/c^2 + (\Delta m)L = 0$$

which yields

$$\boxed{\Delta m = \Delta E/c^2} \tag{27}$$

This equation asserts that energy has mass, that is, energy has inertia.

Among nonscientists, $E = mc^2$ is often regarded as the focal point of the theory of relativity. But the above derivation of Eq. (27) by means of Einstein's *Gedankenexperiment* does not explicitly invoke relativity at all! Relativity enters into this *Gedankenexperiment* only in an indirect and subtle way, because Maxwell's equations, which give us the relation between the energy and the momentum of light, are intrinsically relativistic. In the next section we will see how Eq. (27) fits into relativity.

Although the derivation of Eq. (27) is directly applicable only to the energy of light waves, it has broad implications for other forms of energy. We can take some other form of energy located at the emitter and convert it into light energy by some process. We can then send this light energy to the absorber and finally reconvert it into its original form of energy at the location of the absorber. This sequence of steps effectively shifts the original energy from the left end of the box to the right, and Einstein's *Gedankenexperiment* then tells us that this displacement of the energy is associated with a displacement of mass. Thus, we must associate mass with all forms of energy. Whenever we add an amount ΔE of any kind of energy to a system, we increase its mass by $\Delta E/c^2$. (In the case of potential energy, the situation is somewhat complicated. For instance, if we push an electron from the positive plate of a capacitor to the negative plate, we increase its electric potential energy. However, this potential energy is not actually localized in the electron; it is a joint property of the capacitor-electron system. Thus, what increases is the mass of the capacitor-electron system, not the mass of the electron.)

Equation (27) tells us not only the amount of mass associated with a given amount of energy but also how much energy is made available by the decrease of rest mass in a reaction. A dramatic example of a reaction involving a decrease of rest mass is the annihilation of a particle and its antiparticle, such as the annihilation of an electron (e) and an antielectron ($\bar{\text{e}}$), which results in the creation of two energetic photons, or gamma rays (γ),

$$\text{e} + \bar{\text{e}} \rightarrow \gamma + \gamma \tag{28}$$

The initial rest mass is the sum of the electron and antielectron masses, $m_e + m_e$, or $2m_e$; the photons have no rest mass, so the final rest mass is zero. The decrease of rest mass is therefore $2m_e$. Suppose that the reaction releases an amount of energy ΔE, which is carried away by the photons. The mass associated with this photon energy is $\Delta E/c^2$, and this final mass must match the initial mass $2m_e$; otherwise, the center of mass of the system consisting of the reaction products and the container in which the reaction occurs would fail to remain stationary. Hence $\Delta E/c^2 = 2m_e$, and we find that the energy released in the reaction is

$$\Delta E = 2m_e c^2 = 2 \times 0.511 \text{ MeV} = 1.22 \text{ MeV}$$

This is a large amount of energy. To put it into perspective, imagine that we annihilate a kilogram of electrons with a kilogram of antielectrons (or annihilate a kilogram of any kind of matter with a kilogram of antimatter). The energy released in this annihilation and carried away by the photons is

$$\Delta E = 2 \times 1 \text{ kg} \times c^2 = 1.8 \times 10^{17} \text{ J}$$

If we convert the photons into thermal energy and into electric energy, we have here enough energy to supply the demand of a large city for a year.

The example of annihilation of matter and antimatter demonstrates that the rest mass of a particle is equivalent to an amount mc^2 of energy. Because of this equivalence, it is customary in nuclear and particle physics to adopt a mass unit that is a unit of energy divided by c^2. Thus, the mass of the electron is 0.511 MeV/c^2, the mass of the proton is 938.3 MeV/c^2, and the mass of the neutron is 939.6 MeV/c^2.

Equation (27) is routinely used in nuclear physics to calculate the energies released or absorbed in nuclear reactions and to calculate the binding energies of nuclei. For example, consider the fusion of a hydrogen nucleus (a proton, p) with a deuterium nucleus (D), which produces a nucleus of an isotope of helium (^{3}He) and a photon (γ), according to the reaction

$$\text{p} + \text{D} \rightarrow {}^3\text{He} + \gamma \tag{29}$$

This reaction is one of the fusion reactions that occurs in the core of Sun, where it supplies energy that is ultimately radiated by the Sun in the form of heat and light. The energy released by this reaction can be readily calculated from the masses of the nuclei. The mass of the deuterium nucleus is 1875.6 Mev/c^2, and the mass of the ^{3}He nucleus is 2808.4 MeV/c^2. Hence the decrease of rest mass in the reaction is

$$\Delta m = 938.3 \text{ MeV}/c^2 + 1875.6 \text{ MeV}/c^2 - 2808.4 \text{ MeV}/c^2 = 5.5 \text{ MeV}/c^2$$

Accordingly, the reaction releases an energy $\Delta E = \Delta mc^2 = 5.5$ MeV, which is about 0.2% of the initial rest-mass energy. The energy released in fusion is much smaller than the energy released in the complete annihilation of matter and antimatter, but it is still a large amount of energy. The fusion of a kilogram of proton-deuterium mixture releases 2×10^{14} J. This large energy released in fusion accounts for the Sun's ability to keep on radiating for many millions of years.

Note that although the deuterium nucleus consists of a proton and neutron bound together, its mass of 1875.6 MeV/c^2 is less than the sum of the proton and the neutron masses, 938.3 MeV/c^2 + 939.6 MeV/c^2 = 1876.9 MeV/c^2. The difference of 1.3 MeV/c^2 is the **mass defect** of the deuterium nucleus. This mass defect represents the **binding energy** of the constituents of the nucleus--when a proton and a neutron bind together in a deuterium nucleus, they release an energy of 1.3 MeV, and if we want to separate the deuterium into its constituent proton and neutron, we have to supply this amount of energy. In general, the binding energy (B.E.) of any nucleus can be calculated from its mass defect,

$$\boxed{\text{B.E.} = (\text{mass defect})/c^2} \tag{30}$$

This equation relating the binding energy and the mass defect is also valid for atomic systems--the mass of an atom differs from the sum of the masses of its constituent electrons and nucleus by an amount that represents the binding energy of the electrons. However, electronic binding energies are relatively small, only a few electron-volts, and the resulting mass defects of atoms are too small to be detected by measurements of atomic masses.

Equation (27) is also widely used in high-energy physics, because many of the reactions initiated by collisions among high-energy particles produced in accelerators involve a conversion of rest mass into energy or of energy into rest mass. We will examine a variety of such reactions in the next chapter.

Whenever an amount of rest mass Δm is converted into energy in any process, the resulting amount of energy is $\Delta E = \Delta mc^2$. Since the mass (or the inertia) associated with this energy is $\Delta E/c^2$, which equals Δm, the net amount of mass does not actually change; only the form of the mass changes, from rest mass to "energy mass." Thus, all processes conserve mass, and all processes conserve energy. Conservation of mass is equivalent to conservation of energy. Mass and energy are like two currencies, say U.S. dollars and Hong Kong dollars, that we can use in the calculation of the sum of financial assets held by a bank that has both U.S. dollar accounts and Hong Kong dollar accounts. The relation between these currencies is fixed, 1 Hong Kong \$ = 0.129 U.S.\$. We can calculate the "bottom line" in U.S. dollars or in Hong Kong dollars, and we can convert from one to the other by means of the fixed conversion factor 0.129. Likewise, c^2 is the fixed conversion factor from mass to energy.

4.5 THE ENERGY-MOMENTUM VECTOR; PARTICLES OF ZERO MASS

In the preceding section we saw that rest mass is a form of energy, that is, a free particle at rest has an energy mc^2, called its **rest energy**. The **total energy** of a moving particle is then the sum of the rest energy and the kinetic energy,

$$E = mc^2 + K = mc^2 + \frac{mc^2}{\sqrt{1 - v^2/c^2}} - mc^2$$

or

$$\boxed{E = \frac{mc^2}{\sqrt{1 - v^2/c^2}}} \tag{31}$$

Note that we can express the momentum as

$$\mathbf{p} = \frac{E}{c^2}\mathbf{v} \tag{32}$$

which we can also use to calculate the speed of the particle for a given energy and momentum,

$$v = c^2 \frac{p}{E} \tag{33}$$

If we recall the expression for the proper time interval, $d\tau = dt\sqrt{1 - v^2/c^2}$, we can write the formula for the total energy in the compact form

$$\boxed{E = mc^2 \frac{dt}{d\tau}} \tag{34}$$

Correspondingly, we can write the formula for the momentum, $\mathbf{p} = m\mathbf{v}/\sqrt{1 - v^2/c^2} = m\, d\mathbf{x}/dt\sqrt{1 - v^2/c^2}$, in the compact form

$$\boxed{\mathbf{p} = m\,\frac{d\mathbf{x}}{d\tau}} \tag{35}$$

The important advantage of writing the energy and the momentum in the forms (34) and (35) is that by inspection of these expressions, we can easily discover the Lorentz transformation equations for energy and momentum. The denominator $d\tau$ in these equations is invariant under Lorentz transformations. We can recognize this by recalling that $d\tau$ is the proper time, or the time measured by a clock carried with the particle; this is obviously independent of the laboratory and the spaceship reference frames, and it is not affected by a transformation from one of these reference frames to the other. Alternatively, we can recognize the invariance of $d\tau$ by including the factor dt in the square root:

$$\begin{aligned} d\tau = dt\sqrt{1 - v^2/c^2} &= \frac{1}{c}\sqrt{(dt)^2[c^2 - (dx/dt)^2 - (dy/dt)^2 - (dz/dt)^2]} \\ &= \frac{1}{c}\sqrt{c^2(dt)^2 - (dx)^2 - (dy)^2 - (dz)^2} \end{aligned}$$

Here, the quantity under the integral sign is the square of the spacetime interval, which we know to be invariant under Lorentz transformations.

Since the denominators in Eqs. (34) and (35) are invariant, the Lorentz transformation affects only the numerators, which are proportional to dt and to dx, dy, dz, respectively,

$$E/c^2 \propto dt, \qquad p_x \propto dx, \qquad p_y \propto dy, \qquad p_z \propto dz$$

This means that the Lorentz transformation equations for E/c^2 and the momentum components p_x, p_y, p_z are those for dt and for dx, dy, dz, that is, they are simply those for t, x, y, z:

$$E'/c^2 = \frac{E/c^2 - Vp_x/c^2}{\sqrt{1 - V^2/c^2}}$$

or

$$\boxed{E' = \frac{E - Vp_x}{\sqrt{1 - V^2/c^2}}} \tag{36}$$

and

$$p'_x = \frac{p_x - VE/c^2}{\sqrt{1 - V^2/c^2}} \tag{37}$$

$$p'_y = p_y \tag{38}$$

$$p'_z = p_z \tag{39}$$

In Chapter 2 we saw that ct,x,y,z can be regarded as the components of a four-dimensional position vector spacetime. Likewise, E/c, p_x, p_y, p_z can be regarded as the components of a four-dimensional vector in spacetime called the **energy-momentum vector**. The components of this vector can be written in the alternative forms

$$E/c = \frac{mc}{\sqrt{1 - v^2/c^2}}$$

$$p_x = \frac{mv_x}{\sqrt{1 - v^2/c^2}}, \quad p_y = \frac{mv_y}{\sqrt{1 - v^2/c^2}}, \quad p_z = \frac{mv_z}{\sqrt{1 - v^2/c^2}} \tag{40}$$

and

$$E/c = mc\frac{dt}{d\tau}, \quad p_x = m\frac{dx}{d\tau}, \quad p_y = m\frac{dy}{d\tau}, \quad p_z = m\frac{dz}{d\tau} \tag{41}$$

The Lorentz transformations (36) and (37) mix energy and momentum, just as the Lorentz transformations in Chapter 2 mix time and space. An immediate consequence of this mixing is that the conservation laws for energy and momentum are not independent--we cannot have conservation of one without conservation of the other. Since we took conservation of momentum as the starting point of our discussion of relativistic dynamics, let us see what the conservation of momentum implies for the conservation of energy.

The changes of energy and momentum suffered by a particle in a collision obey the same Lorentz transformation equations as the quantities themselves, and the sums $\Sigma\Delta E$ and $\Sigma\Delta p_x$ of all the changes of energy and momentum for all the particles in the system also obey these Lorentz transformation equations:

$$\Sigma\Delta E' = \frac{\Sigma\Delta E - V\Sigma\Delta p_x}{\sqrt{1 - V^2/c^2}} \tag{42}$$

$$\Sigma \Delta p'_x = \frac{\Sigma \Delta p_x - V \Sigma \Delta E/c^2}{\sqrt{1 - V^2/c^2}} \tag{43}$$

The momentum is conserved in both reference frames, so $\Sigma \Delta p'_x = 0$ and $\Sigma \Delta p_x = 0$. Equation (43) then tells us that $\Sigma \Delta E = 0$, which is the statement of energy conservation in the laboratory frame. With this Eq. (42) next tells us that $\Sigma \Delta E' = 0$, which is the statement of energy conservation in the spaceship frame. We have therefore deduced conservation of energy from conservation of momentum, which is something we cannot do in Newtonian mechanics.

Another remarkable aspect of this deduction is that it does *not* depend on any assumption about the elasticity of the collisions.* The total relativistic energy is conserved even in inelastic collisions. The difference between elastic and inelastic collisions is not in the behavior of the total energy, which is always conserved, but in the behavior of the kinetic and rest-mass energies. In an elastic collision, the kinetic energy and the rest-mass energy are *separately* conserved. But in an inelastic collision, the kinetic energy and the rest-mass energy are not separately conserved. Some kinetic energy is converted into rest mass, or some rest mass into kinetic energy, and only the sum of the kinetic energy and the rest-mass energy is conserved. The particle annihilation reaction (28) is an example of such a conversion of rest-mass energy into "kinetic" energy of the photons. In the next chapter we will analyze other examples of inelastic reactions.

In the case of the four-component position vector, we constructed an invariant from the squares of the components, that is, the square of the spacetime interval, $c^2t^2 - x^2 - y^2 - z^2$. Likewise, in the case of the energy-momentum vector, we can construct the invariant $E^2/c^2 - p_x^{\ 2} - p_y^{\ 2} - p_z^{\ 2} = E^2/c^2 - p^2$. If we substitute the expressions for E and p from Eqs. (31) and (8), we find that the value of this invariant is proportional to the square of the rest mass,

$$E^2/c^2 - p^2 = \frac{m^2c^2}{1 - v^2/c^2} - \frac{m^2v^2}{1 - v^2/c^2} = m^2c^2 \tag{44}$$

This can also be rewritten as

$$\boxed{E^2 = c^2p^2 + m^2c^4} \tag{45}$$

* Since our derivation of the relativistic momentum in Section 4.2 was based on the examination of an elastic collision, it would seem that all our previous arguments depend on elasticity. But this is not the case. We can deduce the same expression for the relativistic momentum if, instead of the elastic collision in Section 4.2, we consider a totally inelastic collision; see Problem 3.

which is a convenient formula for calculating the momentum from the energy, or vice versa.

Table 4.3 summarizes the analogies between the four-component position vector in spacetime and the energy-momentum vector.

TABLE 4.3 THE POSITION VECTOR AND THE ENERGY-MOMENTUM VECTOR

	Position Vector	Energy-Momentum Vector
t component	ct	E/c
x component	x	p_x
y component	y	p_y
z component	z	p_z
Invariant	$c^2t^2 - x^2 - y^2 - z^2$	$E^2/c^2 - p_x{}^2 - p_y{}^2 - p_z{}^2$

Note that in a plot of energy vs. momentum, Eq. (45) is the equation of a hyperbola or, with two space dimensions, the equation of a hyperboloid. This hyperboloid is called the **mass shell**. Energy-momentum vectors that are consistent with Eq. (45) can be represented as vectors from the origin to a point on the mass shell (Fig. 4.9).

Equation (45) is also applicable to particles of zero mass, such as photons and neutrinos (according to some recent experiments, the neutrino maybe has a small mass, but we will ignore that here). For this special case, Eq. (45) reduces to

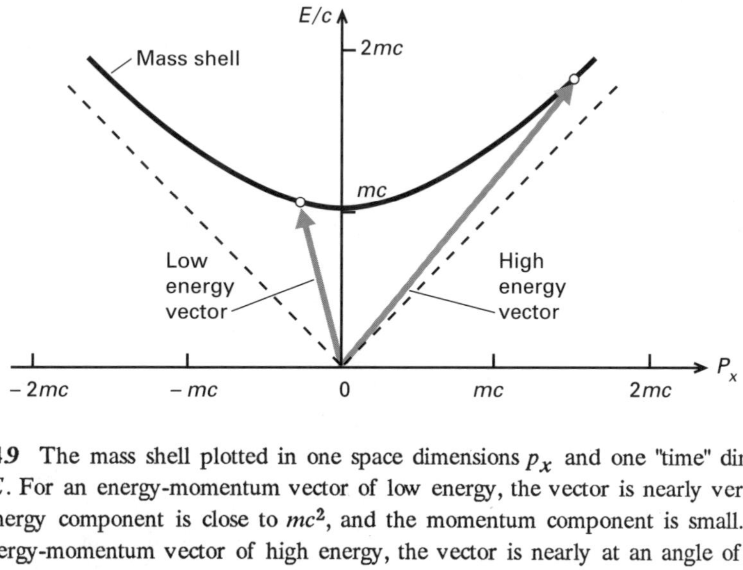

Fig. 4.9 The mass shell plotted in one space dimensions p_x and one "time" dimension E. For an energy-momentum vector of low energy, the vector is nearly vertical; the energy component is close to mc^2, and the momentum component is small. For an energy-momentum vector of high energy, the vector is nearly at an angle of 45^0; the energy is much larger than mc^2, and the momentum component is nearly of the same magnitude as the energy component.

$$\boxed{E = cp} \tag{46}$$

Thus, for a zero-mass particle, the energy is proportional to the momentum. Not surprisingly, this relationship is well known to hold for plane electromagnetic waves, which consist of a stream of photons of parallel momenta. But the relationship (46) is also approximately valid for any particle of very high energy, that is, energy much larger than mc^2. For such an **ultrarelativistic particle**, the mass term in (45) can be neglected compared with the energy term, and the result is (46).

For ultrarelativistic particles, the energy and the momentum differ only by a factor of c, and it is therefore convenient to adopt a momentum unit that is equal to the energy unit divided by c. Thus, high-energy physicists often use GeV/c or TeV/c as a momentum unit. For instance, according to Eq. (46), an electron of energy $E = 50$ GeV has a momentum of approximately $p \simeq E/c = 50$ GeV/c.

By comparing Eqs. (33) and (46), we see that for a particle of zero mass, the speed must always be exactly equal to the speed of light. This prediction of relativity has been tested experimentally for photons and for neutrinos. The most precise observational evidence comes from observations of pulses of photons and neutrinos emitted by distant supernovas and distant pulsars. The available data on the supernova SN1987A establishes that the neutrino pulse and the light pulse arrived within a few hours of each other after traveling a distance of 1.6×10^5 light-years. This simultaneity of arrival requires that the speeds of the neutrinos and the photons agree to within 2 parts in 10^9 [12].

Tighter limits have been extracted from the observed periodic arrivals of high-energy particles emitted by the pulsar Hercules X-1. Some of these particles have energies of a few hundred TeV, some have energies of a few thousand TeV, and they produce large showers of secondary particles (debris) when they strike the upper layers of the atmosphere. The particles are either neutrinos or photons, and they arrive on a fixed schedule, at predictable times, in step with the pulsations of the pulsar. (However, the particles often skip pulsations; they behave like a traveler who can take a train into the city at a fixed time each day if he wants to but elects to go only once every few days.) The fixed schedule of arrival times indicates that all the particles take the same time to complete the trip, even though their energies are different. The observational data indicate that the particles' speeds are identical to within 2 parts in 10^{13} [13].

For photons, the tightest limits are set by observations of pulses from the Crab pulsar and by observations of pulses from gamma-ray bursters. These pulses contain photons of different energies, but since they arrive together, they must have the same speeds. The Crab pulses set a limit of 5 parts in 10^{17} [14], and the gamma-ray pulses set a limit of 6 parts in 10^{21} on the variation of the photon speed with energy [15].

4.6 MOTION IN UNIFORM ELECTRIC AND MAGNETIC FIELDS

Here we will solve the relativistic equation of motion for two special cases of considerable practical importance: motion in a uniform (and constant) electric field and motion in a uniform (and constant) magnetic field.

In a uniform electric field **E**, the force on a charged particle is **F** = q**E**, and the relativistic equation of motion (13) becomes

$$m \frac{d}{dt} \frac{\mathbf{v}}{\sqrt{1 - v^2/c^2}} = q\mathbf{E} \tag{47}$$

For simplicity, we will assume that the initial velocity is in the direction of the electric field, say, the x direction. The motion is then one-dimensional, and the differential equation for the x component of the velocity is*

$$qE = m \frac{d}{dt} \frac{v_x}{\sqrt{1 - v_x{}^2/c^2}} = \frac{m}{\sqrt{1 - v_x{}^2/c^2}} \frac{dv_x}{dt} + mv_x \frac{d}{dt} \frac{1}{\sqrt{1 - v_x{}^2/c^2}} \tag{48}$$

$$= \frac{m}{\sqrt{1 - v_x{}^2/c^2}} \frac{dv_x}{dt} + mv_x \frac{v_x/c^2}{(1 - v_x{}^2/c^2)^{3/2}} \frac{dv_x}{dt} \tag{49}$$

Combining the two terms on the right side with a common denominator, we obtain

$$\frac{qE}{m} = \frac{1}{(1 - v_x{}^2/c^2)^{3/2}} \frac{dv_x}{dt} \tag{50}$$

This equation coincides with the equation for a particle moving with constant acceleration $a'_x = qE/m$ in its own rest frame [see Eq. (2.44)]. We have already integrated this equation in Section 2.6, with the results

$$\boxed{v_x = \frac{qEt}{m} \frac{1}{\sqrt{1 + (qEt/mc)^2}}} \tag{51}$$

* In this section, E is the magnitude of the electric field, not to be confused with the particle energy E of the preceding section.

$$x = \frac{c^2}{qE/m}\left[\sqrt{1 + (qEt/mc)^2} - 1\right] \tag{52}$$

■ **EXAMPLE 3.** At the Stanford Linear Accelerator, the electrons are accelerated from an initial kinetic energy of about zero to a final energy of 50 GeV along a straight track 3 km in length. Assume that this acceleration is produced by a uniform electric field. What field strength is required? How long does the acceleration take to reach the final energy?

SOLUTION. The work done by the electric field is eEx, and this must equal the energy gain,

$$eEx = 50 \text{ GeV}$$

Hence

$$E = \frac{50 \text{ GeV}}{ex} = \frac{50 \times 10^9 \, e \times \text{volt}}{ex} = \frac{50 \times 10^9 \text{ volt}}{3000 \text{ m}} = 1.7 \times 10^7 \text{ volt/m}$$

From Eq. (51) we see that qEt/mc must be large to achieve a speed close to the speed of light. Taking this into account in Eq. (52), we find approximately

$$x \simeq \frac{c^2}{qE/m}\frac{qEt}{mc} = ct \tag{53}$$

This merely says that the electrons achieve a speed close to c almost immediately and travel at a speed close to c almost the full length of the track. The time taken for the complete run is then

$$t \simeq \frac{x}{c} = \frac{3000 \text{ m}}{3 \times 10^8 \text{ m/s}} = 1 \times 10^{-5} \text{ s}$$

In a uniform magnetic field **B**, the force on a charged particle is $q\,\mathbf{v} \times \mathbf{B}$, and the relativistic equation of motion is

$$m\frac{d}{dt}\frac{\mathbf{v}}{\sqrt{1 - v^2/c^2}} = q\,\mathbf{v} \times \mathbf{B} \tag{54}$$

Since the force is perpendicular to the velocity, it does no work on the particle, and the kinetic energy remains constant. This means that the magnitude of the velocity remains constant, and only its direction changes. The result is uniform circular motion along a circle of radius r, as in the Newtonian case. The centripetal acceleration has the familiar value $|d\mathbf{v}/dt| = v^2/r$ (this expression for the centripetal acceleration depends only on the geometry of circular motion and is equally valid for relativistic and Newtonian motions). With a constant value of v^2, Eq. (54) then reduces to

$$\frac{m}{\sqrt{1 - v^2/c^2}} \frac{v^2}{r} = qvB$$

or

$$r = \frac{1}{qB} \frac{mv}{\sqrt{1 - v^2/c^2}} \tag{55}$$

In terms of the relativistic momentum p, this becomes

$$\boxed{r = \frac{p}{qB}} \tag{56}$$

This simple relation between the momentum and the radius of the motion is of the same form as in the Newtonian case, but the momentum in (56) is the relativistic momentum. The relation (56) is routinely used in high-energy particle physics to determine the momentum of charged particles from the observed radius of their orbit in a known, uniform magnetic field.

The frequency of the circular motion, called the **cyclotron frequency**, is

$$\nu = \frac{v}{2\pi r} = \frac{qB}{2\pi m} \sqrt{1 - v^2/c^2} \tag{57}$$

This is smaller than the Newtonian value $qB/2\pi m$ by the factor $\sqrt{1 - v^2/c^2}$.

The velocity dependence of the relativistic cyclotron frequency complicates the operation of **cyclotrons**, where particles in a circular orbit in a magnetic field are accelerated to higher and higher energies by means of repetitive pushes from a tangential electric field. At low energies, the frequency is constant, and the pushes can be applied periodically. But at high energies, the cyclotron frequency decreases according to Eq. (57), and the pushes must be applied more and more slowly. This means that the particles in such a cyclotron must be accelerated in bunches, the timing of the pushes being matched to the orbital motion of the bunch. Cyclotrons that operate with a gradually decreasing frequency matched to bunches of particles are called synchrocyclotrons. Alternatively, it is possible to increase the strength of the magnetic field gradually in such a way that the product $B\sqrt{1 - v^2/c^2}$ remains constant during the acceleration of a bunch of particles. Cyclotrons that operate with such a gradually increasing magnetic field are called synchrotrons.

Most of the particle accelerators used in high-energy research are synchrotrons; among these, the largest are the CERN LEP accelerator (Fig. 4.6) and the Fermilab accelerator (Figs. 4.7 and 5.4). Most accelerators used for cancer therapy also are synchrotrons. Such medical accelerators produce beams of protons that either are used directly for irradiation of tumors or are used to generate secondary beams of neutrons, by impact of the protons on a target (Fig. 4.10). Some medical accelerators are linear; these produce beams of electrons that are used to generate secondary beams of photons, that is X rays of extremely high energies.

Fig. 4.10 A medical accelerator at the University of Washington Medical Center in Seattle. (H. G. Blosser, NSCL, Michigan State University)

REFERENCES

1. C. T. Zahn and A. H. Spees, Phys. Rev. **53**, 511 (1938).
2. P. S. Faragó and L. Jánossy, Nuovo Cimento **5**, 1411 (1957).
3. S. Parker, Am. J. Phys. **40**, 241 (1972).
4. K. N. Geller and R. Kollarits, Am. J. Phys. **40**, 1125 (1972).
5. D. J. Grove and J. G. Fox, Phys. Rev. **90**, 378 (1953).
6. V. P. Zrelov, A. A. Tyapkin, and P. S. Farago, Sov. Phys. JETP **7**, 384 (1958).
7. B. C. Brown, G. E. Masek, T. Maung, E. S. Miller, H. Ruderman, and W. Vernon, Phys. Rev. Lett. **30**, 763 (1973).
8. Z. G. T. Guiragossian, G. B. Rothbart, M. R. Yearian, R. A. Gearhart, and J. J. Murray, Phys. Rev. Lett. **34**, 335 (1975).
9. D. Newman, G. W. Ford, A. Rich, and E. Sweetman, Phys. Rev. Lett. **40**, 1355 (1978).

10. R. Van Dyck, P. Schwinberg, and H. Dehmelt, Phys. Rev. Lett. **38**, 310 (1977).
11. J. Wesley and A. Rich, Phys Rev. *A* **4**, 1341 (1971).
12. M. J. Longo, Phys. Rev. *D* **36**, 3276 (1987)
13. T. J. Haines, D. E. Alexandreas, R. C. Allen, S. Biller, D. Berley, R. L. Burman, D. R. Cady, C. Y. Chang, B. L. Dingus, G. M. Dion, R. W. Ellsworth, J. A. Goodman, C. M. Hoffman, J. Lloyd-Evans, D. E. Nagle, M. Potter, V. D., Sandberg, C. A. Wilkinson, and G. B. Yohd, Phys. Rev. *D* **41**, 692 (1990).
14. B. Warner and R. Nather, Nature **222** 157 (1969).
15. B. E. Schaefer, Phys. Rev. Lett. **82**, 4964 (1999).

PROBLEMS

1. From the relativistic equation of motion (13), prove that if $\mathbf{F} = 0$, then $d\mathbf{v}/dt = 0$.

2. (a) Show that the relativistic equation of motion (13) can be put in the form

$$m\frac{d\mathbf{v}}{dt} = \sqrt{1 - v^2/c^2}\,(\mathbf{F} - \mathbf{v}\ \mathbf{F}\cdot\mathbf{v}/c^2)$$

(b) Show that if we define "transverse" and "longitudinal" masses

$$m_{\mathrm{T}} = \frac{m}{\sqrt{1 - v^2/c^2}} \quad \text{and} \quad m_{\mathrm{L}} = \frac{m}{(1 - v^2/c^2)^{3/2}}$$

then the equation of motion can be put in the alternative form

$$\frac{d\mathbf{v}}{dt} = \frac{\mathbf{F}_{\mathrm{T}}}{m_{\mathrm{T}}} + \frac{\mathbf{F}_{\mathrm{L}}}{m_{\mathrm{L}}}$$

where $\mathbf{F}_{\mathrm{T}}$ is the component of the force transverse to the velocity, and $\mathbf{F}_{\mathrm{L}}$ is the component of the force parallel (or longitudinal) to the velocity.
(c) Show that if both $\mathbf{F}_{\mathrm{T}}$ and $\mathbf{F}_{\mathrm{L}}$ are nonzero and if v is near c, then the direction of the acceleration is nearly perpendicular to the velocity and does not depend on the direction of the force. What is the maximum possible angle between the force and the acceleration?

3. In Section 4.2 we used a special elastic collision between two particles to derive the relativistic expression for the momentum. Repeat this derivation if the collision is totally inelastic instead of elastic. Remember that in a totally inelastic collision, the two particles stick together, and they form a single particle of a momentum equal to the sum of the individual momenta before the collision.

4. Calculate the speeds of protons of kinetic energies 0.50 GeV, 1.00 GeV, 2.00 GeV, and 10.00 GeV, respectively.

5. In a hydrogen atom, the electron has an orbital speed of 2.2×10^6 m/s. For this speed, what is the fractional difference between the relativistic and Newtonian expressions for kinetic energy?

6. The orbital speed of the Sun around the center of the Galaxy is about 250 km/s. For this speed, what is the fractional difference between the relativistic and Newtonian expressions for kinetic energy? for momentum?

7. The Stanford Linear Accelerator produces electrons of energy 50 GeV and a speed $0.999999999948c$ (see Example 3). What would be the energy of a proton of the same speed?

8. Prove that the relation between the kinetic energy K and the momentum p can be expressed in the alternative forms

$$\frac{p^2}{K} = 2m + \frac{K}{c^2} \quad \text{and} \quad cp = \sqrt{K(K + 2mc^2)}$$

The first of these shows that the plot of p^2/K vs. K is a straight line. Such a plot can be used to analyze experimental data of measurements of K and p; deviations from the relativistic formulas would show up as deviations from the straight line.

9. The SSC accelerator, which was partially built and then abandoned, was supposed to accelerate protons to an energy of 20 TeV. How close to the speed of light is the speed of a proton of this energy? What would be the energy of an electron of the same speed?

10. We want to send a space probe on an interstellar mission to a nearby star. The space probe is to have a speed of $0.5c$.
(a) Suppose the probe has a mass of 100 kg. What is its kinetic energy? Compare this kinetic energy with the yearly energy expenditure of the United States, about 10^{20} J.
(b) Suppose the probe is miniaturized and has a mass of only 1 g. What is its kinetic energy in this case?

11. The Sun is radiating heat and light at the rate of 3.9×10^{26} W. In 1 year, how much mass does the Sun lose because of this radiation?

12. The energy of nuclear explosions is usually rated in kilotons or megatons TNT, and 1 kiloton TNT corresponds to an energy release of 4.2×10^{12} J. Suppose that a 10-kiloton nuclear device is set off in an underground explosion.
(a) How much rest mass of nuclei is converted into energy in this explosion?
(b) Assuming that all of the energy remains trapped underground, is there a change of the total mass of the Earth?

13. The hydrogen gas at the center of the Sun has a temperature of 1.5×10^7 K. What is the thermal energy per kilogram of this gas? What is the mass associated

with this kinetic energy? Compare this extra mass with the rest mass of the gas, and find the percentage mass increment attributable to the kinetic energy.

14. The strongest magnetic fields produced in laboratories, by explosive compression of magnetic field lines, is 10^3 T. In such a magnetic field, what is the energy density and what is the mass density?

15. Calculate the mass defect of the ^{3}He nucleus, which consists of two protons and one neutron bound together.

16. The electron in the ground state of the hydrogen atom has a binding energy of 13.6 eV. What is the corresponding mass defect of the hydrogen atom? Express this as a percentage of the mass of the atom.

17. The helium nucleus ^{4}He consists of two protons and two neutrons bound together. The mass of this nucleus is 3727.4 MeV$/c^2$.
(a) Calculate the binding energy of this nucleus.
(b) From Section 4.4, we know that the binding energy of the deuterium nucleus is 1.3 MeV. How much energy is released when two deuterium nuclei are allowed to fuse into a helium nucleus, according to the reaction $D + D \rightarrow {}^4He$?

18. The bombardment of lithium by protons leads to a disintegration of the lithium nucleus into two helium nuclei, according to the reaction $p + {}^7Li \rightarrow {}^4He + {}^4He$. This reaction is of some historical interest, because it was the first nuclear reaction induced with artificially accelerated protons from the electrostatic accelerator of Cockcroft and Walton (1932). The mass of the ^{7}Li nucleus is 6533.8 MeV$/c^2$. Calculate the kinetic energy of each helium nucleus if the incident proton has nearly zero kinetic energy. Compare this with the experimentally measured value of 8.68 MeV.

19. While moving at a speed of $0.90c$ relative to our laboratory, a $K^0{}_S$ meson decays into two pions, each of mass 140 MeV$/c^2$ and kinetic energy 110 MeV. In the rest frame of the K meson, one of these pion moves in the forward direction and the other in the backward direction. What are the kinetic energy and the momentum of each pion in the laboratory frame?

20. For a zero-mass particle moving along the positive or negative x axis, $p_x = \pm p = \pm E/c$. Show that in this case, the two transformation laws (36) and (37) coincidc, and that each givcs

$$E' = E\sqrt{\frac{1 \mp V/c}{1 \pm V/c}}$$

Explain why this result is similar to the formulas for the relativistic Doppler shift, Eqs. (3.18) and (3.19).

21. The phase of a light wave of angular frequency ω and wave vector $\mathbf{k}$ is defined as $\omega t - k_x x - k_y y - k_z z$, where k_x, k_y, k_z are the components of the wave vector. The phase determines where the wave has minima or maxima, and it must therefore be invariant under Lorentz transformations. From this condition of the invariance of the phase, deduce the Lorentz transformation equations for ω, k_x, k_y, and k_z. Compare your result with the transformation equations for energy and momentum. (*Hint*: Begin with $\omega t - k_x x - k_y y - k_z z = \omega' t' - k'_x x' - k'_y y' - k'_z z'$, and express t', x', y', z' in terms of t, x, y, z. Then take the partial derivative of the equation with respect to t, or x, or y, or z.)

22. Consider two energy-momentum vectors with components $E_1/c, p_{1x}, p_{1y}, p_{1z}$ and $E_2/c, p_{2x}, p_{2y}, p_{2z}$, respectively. Show that the "scalar product" $E_1 E_2/c^2 - p_{1x}p_{2x} - p_{1y}p_{2y} - p_{1z}p_{2z}$ is invariant under Lorentz transformations.

23. Show that the quantity $Et - p_x x - p_y y - p_z z$, or $Et - \mathbf{p} \cdot \mathbf{x}$, is invariant under Lorentz transformations. This quantity plays an important role in quantum mechanics, where the wavefunction of a free particle is a function of this quantity.

24. In the Fermilab Tevatron accelerator, protons of kinetic energy 1000 GeV move in a circle of radius 1 km. The protons are held in this circular orbit by a vertical magnetic field. What is the momentum of such protons? Assuming that the magnetic field is uniform, what must be the field strength?

25. Suppose that in a region of space there is a uniform electric field $\mathbf{E}$ and also a uniform electric field $\mathbf{B}$ at right angles to this electric field; such fields are called **crossed fields**. Show that a charged particle entering this region with a velocity $\mathbf{v} = \mathbf{E} \times B/B^2$ at right angles to both $\mathbf{E}$ and $\mathbf{B}$ will pass through this region without any deflection. Show that this result is equally valid in Newtonian and relativistic mechanics. The crossed fields act as a velocity filter: when a beam of charged particles of different velocities enters such a region, the particles with the preferred velocity will pass straight through; particles with other velocities will be deflected out of the beam. What happens if you attempt to operate such a velocity filter with an electric field $E > B$?

26. The Earth's magnetic field at the equator is 0.33×10^{-4} T, and its direction is north. What must be the momentum and the energy of an electron that orbits the Earth above the atmosphere in an equatorial circular orbit of radius 6.5×10^3 km? What is the speed of such an electron, and what is the period of the orbit?

27. A bubble chamber, used to make the tracks of protons and other charged particles visible, is placed between the poles of a large electromagnet that produces a uniform magnetic field of 20 T. In this field, a proton makes a track that is an arc of a circle of radius 3.2 m. What is the momentum of the proton? What is its kinetic energy? What is its speed?

28. In the Crab nebula (the remnant of a supernova explosion), electrons orbit in a magnetic field of about 10^{-8} T. If the momentum of these electrons is about 10^{-16}

kg·m/s, what is the cyclotron frequency? What is the orbital radius?

29. A bubble chamber is placed in an uniform magnetic field of 1.0 T. An electron orbiting in this bubble chamber is observed to have a spiral orbit, because the electron gradually loses energy by interactions with the atoms of the liquid filling the chamber. Suppose that the initial radius of the orbit is 8.0 cm and that a few revolutions later, the radius of the orbit is 1.0 cm. What is the decrease of the energy of the electron?

30. In a small synchrotron, the electrons are held in a circular orbit of radius 15 cm. If the electrons are to be accelerated from an initial kinetic energy of 2 MeV to a final energy of 40 MeV, what must be the initial magnetic field and the final magnetic field? What must be the initial frequency and the final frequency of the tangential electric field that provides the acceleration? Explain why these frequencies are nearly the same.

31. In the laboratory frame, a straight rod is at rest along the x axis. You grasp the two ends of a rod in your two hands and, by suddenly pushing inward at both ends *simultaneously*, you place the rod under compressional stress, and you hold it, without giving it a translational motion. In the spaceship frame the rod has a velocity- V in the negative x' direction, and the pushes of your two hands are not simultaneous. Your push on the left end starts later than that on the right (Leading Late!), and therefore the rod acquires an extra amount of momentum in the negative x' direction. This would seem to indicate that the rod does not remain at rest in the laboratory frame. Resolve this apparent paradox by showing that for a given speed V, the compressionally stressed rod has a momentum larger than an unstressed rod. (*Hint*: In the spaceship frame the rod moves toward the left at speed V, and the push F of your hand does positive work at the rate VF on the right end of the rod, whereas your other hand does negative work at the rate $-VF$ on the left end of the rod. Associated with this work, there is an energy flow toward the left, a mass flow, and a momentum density.)*

---

* I am indebted to R. Lichtenstein for this paradox.

5

Collisions and Conservation of Energy-momentum

I ADHERED SCRUPULOUSLY TO THE PRECEPT OF THAT BRILLIANT PHYSICIST L. BOLTZMANN, ACCORDING TO WHOM MATTERS OF ELEGANCE SHOULD BE LEFT TO THE TAILOR AND THE COBBLER.

Albert Einstein, *Relativity.*

The main application of relativistic mechanics is in the motion of atomic and subatomic particles. This is in contrast to Newtonian mechanics, where we study not only the motion of particles but also the motion of solid bodies, fluids, and a variety of composite systems consisting of several solid bodies linked by springs, strings, flexible rods, etc. The emphasis on particles in the application of relativistic mechanics is motivated in part by practical considerations and in part by theoretical considerations--we have no practical means for accelerating macroscopic bodies to relativistic speeds, and the theoretical treatment of macroscopic bodies performing a combination of translational and rotational motions at relativistic speeds is prohibitively complicated.

In this chapter we will deal with only the simplest cases of particle motion. We will assume that the particles are free, except for brief interactions during collisions. As we saw in the preceding chapter, Newton's Third Law is valid for relativistic particles that interact by contact forces. Hence the total momentum is conserved in collisions, so the total momentum before the collision is equal to the momentum after the collision. From the discussion of the energy-momentum four vector in Section 4.5, we know that this conservation of the total momentum implies a corresponding conservation of the total relativistic energy. We will exploit these conservation laws in this chapter.

As in Newtonian mechanics, a collision is said to be **elastic** if the total kinetic energy before the collision is equal to the total kinetic energy after the collision, and the collision is **inelastic** if the total kinetic energy changes. We will first discuss elastic collisions and then inelastic collisions.

5.1 ELASTIC COLLISIONS

Relativistic elastic collisions can be treated in the same way as nonrelativistic collisions, although the calculations are somewhat more complicated. In the special case of an elastic collision between two particles moving along a straight line, the conservation laws for energy and momentum completely determine the final motion in terms of the initial motion.

Suppose that a moving particle of mass m_1, initial energy E_1, and initial momentum p_1 is incident on an initially stationary particle of mass m_2. After the elastic collision, both particles move along the original line of motion, with momenta $\bar{p}_1$ and $\bar{p}_2$, and energies $\bar{E}_1$ and $\bar{E}_2$, respectively (see Fig. 5.1). The sum of the initial energies is $E_1 + m_2c^2$, and the sum of the final energies is $\bar{E}_1 + \bar{E}_2 = \sqrt{\bar{p}_1^2c^2 + m_1^2c^4} + \sqrt{\bar{p}_2^2c^2 + m_2^2c^4}$. Conservation of energy requires

$$E_1 + m_2c^2 = \sqrt{\bar{p}_1^2c^2 + m_1^2c^4} + \sqrt{\bar{p}_2^2c^2 + m_2^2c^4} \tag{1}$$

and conservation of momentum requires

$$p_1 = \bar{p}_1 + \bar{p}_2 \tag{2}$$

These are two equations for the two unknowns $\bar{p}_1$ and $\bar{p}_2$. To solve these equations, we first use Eq. (2) to eliminate $\bar{p}_2$ from Eq. (1):

$$E_1 + m_2c^2 = \sqrt{\bar{p}_1^2c^2 + m_1^2c^4} + \sqrt{(p_1 - \bar{p}_1)^2c^2 + m_2^2c^4} \tag{3}$$

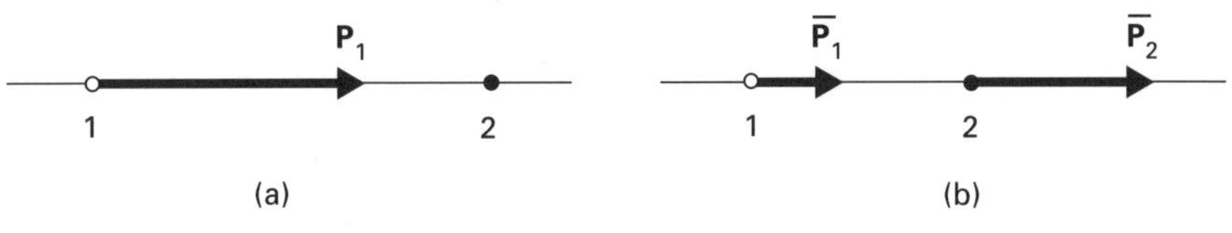

Fig. 5.1 An elastic collision between an initially moving particle 1 and an initially stationary particle 2. The particles move along a straight line. (a) The initial momentum of particle 1 is p_1. (b) The final momenta of particles 1 and 2 are $\bar{p}_1$ and $\bar{p}_2$.

We can rearrange the terms of this equation,

$$E_1 + m_2c^2 - \sqrt{\bar{p}_1^{\,2}c^2 + m_1^{\,2}c^4} = \sqrt{(p_1 - \bar{p}_1)^2c^2 + m_2^{\,2}c^4} \tag{4}$$

and then square both sides:

$$(E_1 + m_2c^2)^2 - 2(E_1 + m_2c^2)\sqrt{\bar{p}_1^{\,2}c^2 + m_1^{\,2}c^4} + \bar{p}_1^{\,2}c^2 + m_1^{\,2}c^4 = (p_1 - \bar{p}_1)^2c^2 + m_2^{\,2}c^4 \tag{5}$$

We can now rearrange again,

$$-2(E_1 + m_2c^2)\sqrt{\bar{p}_1^{\,2}c^2 + m_1^{\,2}c^4} = -(E_1 + m_2c^2)^2 - m_1^{\,2}c^4 + p_1^{\,2}c^2 - 2p_1\bar{p}_1c^2 + m_2^{\,2}c^4 \tag{6}$$

and square both sides again:

$$4(E_1 + m_2c^2)^2(\bar{p}_1^{\,2}c^2 + m_1^{\,2}c^4) = [(E_1 + m_2c^2)^2 + m_1^{\,2}c^4 - p_1^{\,2}c^2 - m_2^{\,2}c^4]^2 + 4p_1\bar{p}_1c^2[(E_1 + m_2c^2)^2 + m_1^{\,2}c^4 - p_1^{\,2}c^2 - m_2^{\,2}c^4] + 4p_1^{\,2}\bar{p}_1^{\,2}c^4 \tag{7}$$

This is a quadratic equation for the unknown $\bar{p}_1$, and its solution is given by the standard formula for such quadratic equations. After some simplifications, we obtain the final result

$$\bar{p}_1 = p_1 \frac{m_1^{\,2} + m_2E_1/c^2 - m_2\sqrt{m_2^{\,2} + m_1^{\,2} + 2m_2E_1/c^2 + p_1^{\,2}/c^2}}{m_1^{\,2} + m_2^{\,2} + 2m_2E_1/c^2} \tag{8}$$

From this, we immediately find

$$\bar{p}_2 = p_1 - \bar{p}_1$$

$$= p_1 \frac{m_2^{\,2} + m_2E_1/c^2 + m_2\sqrt{m_2^{\,2} + m_1^{\,2} + 2m_2E_1/c^2 + p_1^{\,2}/c^2}}{m_1^{\,2} + m_2^{\,2} + 2m_2E_1/c^2} \tag{9}$$

If the motion is nonrelativistic, with $E_1 \simeq m_1c^2$ and $p_1 << m_1c$, these results reduce to

$$\bar{p}_1 = \frac{m_1 - m_2}{m_1 + m_2} p_1 \tag{10}$$

and

$$\bar{p}_2 = \frac{2m_2}{m_1 + m_2} p_1 \tag{11}$$

These are the familiar Newtonian results for an elastic collision between particles moving on a straight line. Note that we have expressed Eq. (10), Eq. (11), and all other preceding equations in terms of momentum, rather than velocity, as is usual in Newtonian mechanics. This is because an attempt to use velocity as the unknown in the relativistic equations would have led to an extremely messy equation of the fourth order. As a general rule, it is best to avoid the use of velocities in problems involving conservation of the relativistic energy and momentum.

One interesting consequence of Eq. (8) is that $\bar{p}_1$ is positive if $m_1 > m_2$ and negative if $m_1 < m_2$ (see Problem 1). Thus, the incident particle continues its forward motion if it has the larger rest mass, and it recoils backward if it has the smaller rest mass. Although this is familiar in Newtonian physics, it seems counterintuitive in relativistic physics, where we might expect that an incident particle of small mass but very large kinetic energy would push the stationary particle out of the way and continue its forward motion, as though it had a large mass. According to Eq. (8), however, the direction of the final motion depends only on the magnitudes of the rest masses, not on the kinetic energy or on the "mass" contributed by the kinetic energy. A corollary of this is that if the incident particle has zero rest mass, it always recoils backward, no matter how large its energy. In this case the stationary particle acquires almost all of the large energy of the incident particle.

The case of an elastic collision on a straight line is exceptional in that the final motion is completely determined by the conservation laws for energy and momentum. In more general cases--such as motion in two or three dimensions and collisions of more than two particles--these conservations laws do not suffice for a complete determination of the final motion. In the next section we will examine in detail the case of an elastic collision between a photon and an electron in two dimensions, and we will see what constraints the conservation laws impose between the final energies and the directions of motion of the particles.

5.2 ENERGY AND MOMENTUM OF THE PHOTON; THE COMPTON EFFECT

Photons, or quanta of light, are particles of zero rest mass. As discussed in Section 4.5, such particles necessarily have a speed equal to the speed of light, and their energy and momentum are directly proportional [see Eq. (4.46)],

$$E = cp \tag{12}$$

From quantum theory we know that the photon is both a particle and a wave. The energy of the photon is related to the frequency by the equation discovered by Planck,

$$E = h\nu \tag{13}$$

and the momentum is related to the wavelength by deBroglie's equation,

$$p = h/\lambda \tag{14}$$

Here h is Planck's constant, $h = 6.63 \times 10^{-34}$ J·s. If we insert Eqs. (13) and (14) into Eq. (12), we obtain

$$\nu = c/\lambda \tag{15}$$

which is the usual relation between the frequency and the wavelength for an electromagnetic wave. This verifies that Eqs. (12), (13), and (14) are consistent. We can then also express the momentum of the photon as

$$p = h\nu/c \tag{16}$$

The notions that electromagnetic radiation has particle properties and that thermal radiation is to be regarded as a gas of photons were first proposed by Einstein in 1909 (he was ultimately awarded the Nobel Prize for this discovery, but not for his discovery of relativity). Direct experimental evidence for the particle properties of the photon was found by A. H. Compton in 1922, during experiments involving the bombardment of a graphite target with X rays.

When a high-energy photon strikes an electron initially at rest within the target, the photon recoils with an increase of its wavelength. This **Compton effect** provides evidence for the particlelike behavior of the photon, since it can be understood in terms of an elastic collision between the photon and the electron. The electron recoils when struck by the photon and thereby acquires some energy from the photon. Correspondingly, the photon suffers a loss of energy, which manifests itself as an increase of the wavelength of the photon.

For the analysis of such a collision, we assume that the electron is initially at rest but is free to move (that is, we neglect the initial speed of the electron, and we neglect its binding energy within the target atom; this is a good approximation if the photon energy is large so the electron recoil energy is also large). To find the change of the energy of the photon, we apply the conservation laws for momentum and energy. The initial momentum of the photon is $h\nu/c$, and the final momentum of the photon is $h\bar{\nu}/c$, which makes an angle θ with respect to the initial momentum. The final electron momentum is $\bar{p}$. Figure 5.2a displays these momenta in a vector diagram. Conservation of momentum requires that the initial momentum of the photon equal the vector sum of the final momenta of the photon and the electron. The momentum vectors therefore form a triangle (see Fig. 5.2b), and the law of cosines tells us that

$$\bar{p}^2 = \left(\frac{h\nu}{c}\right)^2 + \left(\frac{h\bar{\nu}}{c}\right)^2 - 2\,\frac{h\nu}{c}\frac{h\bar{\nu}}{c}\cos\theta \tag{17}$$

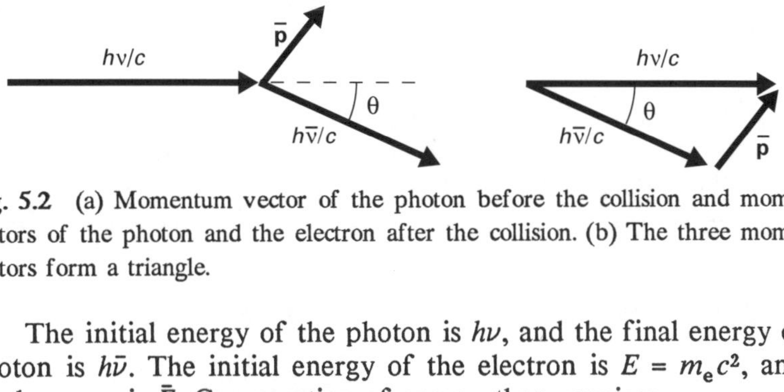

Fig. 5.2 (a) Momentum vector of the photon before the collision and momentum vectors of the photon and the electron after the collision. (b) The three momentum vectors form a triangle.

The initial energy of the photon is $h\nu$, and the final energy of the photon is $h\bar{\nu}$. The initial energy of the electron is $E = m_e c^2$, and the final energy is $\bar{E}$. Conservation of energy then requires

$$h\nu + m_e c^2 = h\bar{\nu} + \bar{E} \tag{18}$$

From this we find that the square of the final electron energy is

$$\bar{E}^2 = (h\nu - h\bar{\nu})^2 + 2m_e c^2(h\nu - h\bar{\nu}) + (m_e c^2)^2 \tag{19}$$

If we subtract c^2 times Eq. (17) from Eq. (19) and use the identity $\bar{E}^2 - c^2\bar{p}^2 = (m_e c^2)^2$, we obtain

$$(m_e c^2)^2 = (h\nu - h\bar{\nu})^2 + 2m_e c^2(h\nu - h\bar{\nu}) + (m_e c^2)^2$$
$$- (h\nu)^2 - (h\bar{\nu})^2 + 2\,h\nu\,h\bar{\nu}\cos\theta$$

or

$$-2\,h\nu\bar{\nu} + 2m_e c^2(\nu - \bar{\nu}) + 2\,h\nu\bar{\nu}\cos\theta = 0$$

Dividing this by $2\nu\bar{\nu}m_e c$, we find

$$-\frac{h}{m_e c} + \frac{c}{\bar{\nu}} - \frac{c}{\nu} + \frac{h}{m_e c}\cos\theta = 0 \qquad (20)$$

Since the wavelength of the photon is $\lambda = c/\nu$, this is an equation for the change of the wavelength:

$$\boxed{\bar{\lambda} - \lambda = \frac{h}{m_e c}(1 - \cos\theta)} \qquad (21)$$

The numerical factor $h/m_e c$ on the right side of this equation is called the **Compton wavelength,**

$$\boxed{\frac{h}{m_e c} = 2.43 \times 10^{-12}\ \text{m}} \qquad (22)$$

The maximum value of the angular factor in Eq. (21) is $1 - \cos\theta = 2$, for photons that recoil in the backward direction. For such photons the change of wavelength is maximum, $\bar{\lambda} - \lambda = 2 \times 2.4 \times 10^{-12}$ m.

5.3 INELASTIC COLLISIONS; NUCLEAR COLLISIONS AND REACTIONS

Relativistic inelastic collisions involve the transformation of kinetic energy into rest-mass energy or of rest-mass energy into kinetic energy. Thus, the rest masses of the particles that emerge from the collision differ from the rest masses of the particles that initiated the collision. For instance, if two atoms engage in an inelastic collision, then one or both of the atoms absorb some of the initial kinetic energy, and correspondingly, their rest mass increases. This is similar to a Newtonian collision between, say, two cannonballs of lead, where some of the initial kinetic energy is expended to produce permanent deformations of the balls, a process that is accompanied by the deposition of a corresponding amount of heat energy in each ball. In collisions between atoms, the relativistic effects are usually small, and the mass changes of the atoms can usually be neglected (although the energy changes cannot).

Experiments on collisions and reactions of nuclei often involve much larger energies than collisions between atoms, and relativistic effects are important. The reactions usually involve the conversion of substantial amounts of kinetic energy into rest-mass energy, or vice versa. Although in most nuclear reactions the nuclei move at nonrelativistic speeds, some of the lighter particles ejected in nuclear reactions--such as electrons, protons, neutrons, and alpha particles--may attain relativistic speeds. Furthermore, any photons and neutrinos involved in a reaction are always fully relativistic.

Inelastic collisions and reactions between nuclei can result not only in a change of the rest masses of the participating nuclei but also in transmutations of some or all of the participating nuclei into other nuclei. Such transmutation reactions can involve a decrease of kinetic energy by conversion into rest mass or an *increase* of kinetic energy by conversion from rest mass. In the former case, the sum of rest masses increases during the reaction (the reaction is endoergic); in the latter case, the sum of rest masses decreases during the reaction (the reaction is exoergic).

Quantitatively, endoergic and exoergic reactions are distinguished by their **Q value**, which is the difference between the sums of rest-mass energies before and after the reaction,

$$\boxed{Q = m_1c^2 + m_2c^2 + \ldots - \bar{m}_1c^2 - \bar{m}_2c^2 - \ldots} \tag{23}$$

Here the overbars indicate the masses after the reaction. If the net kinetic energy is K before the reaction and $\bar{K}$ after, then we can express the conservation of energy as

$$\bar{K} = K + Q \tag{24}$$

For an endoergic reaction, with a negative value of Q, the initial kinetic energy must exceed some least value if the reaction is to be viable. The least energy needed for the viability of the reaction is called the **threshold energy**. Equation (24) suggests that for a negative value of Q, the threshold kinetic energy is $K = |Q|$, since this guarantees a non-negative value of $\bar{K}$. However, as we will see in the next section, the actual threshold is usually larger, because some of the initial kinetic energy is tied up in the motion of the center of mass, and therefore not all of the initial kinetic energy is available for conversion into rest mass.

The first nuclear reaction was discovered by E. Rutherford in 1919, when he bombarded nitrogen nuclei with alpha particles and produced oxygen and protons,

$$\alpha + {}^{14}\mathrm{N} \rightarrow {}^{17}\mathrm{O} + \mathrm{p} \tag{25}$$

This reaction has a negative Q value, and it can proceed only if the alpha particle has a sufficient kinetic energy.

■ **EXAMPLE 1.** Calculate the Q value for the reaction (25). Note that since the alpha particle is a helium nucleus and the proton a hydrogen nucleus, the reaction can also be written as

$$^{4}\text{He} + {}^{14}\text{N} \rightarrow {}^{17}\text{O} + {}^{1}\text{H} \tag{26}$$

The masses of the helium nucleus ^{4}He, the nitrogen nucleus ^{14}N, the oxygen nucleus ^{17}O, and the hydrogen nucleus ^{1}H are $m_{\text{He}} = 3727.4$ MeV/c^2, $m_{\text{N}} = 13{,}040.2$ MeV/c^2, $m_{\text{O}} = 15{,}830.5$ MeV/c^2, and $m_{\text{H}} = 938.3$ MeV/c^2, respectively.*

SOLUTION. With the given masses, the Q value is

$$Q = m_{\text{He}}c^2 + m_{\text{N}}c^2 - m_{\text{O}}c^2 - m_{\text{H}}c^2$$

$$= 3727.4 \text{ MeV} + 13{,}040.2 \text{ MeV} - 15{,}830.5 \text{ MeV} - 938.3 \text{ MeV} = -1.2 \text{ MeV}$$

This confirms that the Q value is negative.

Important reactions with positive Q values are **fission reactions**, in which a heavy isotope splits into two lighter isotopes when bombarded by neutrons. The first such reaction was discovered by O. Hahn and F. Strassmann in 1938, who found that a sample of uranium irradiated with neutrons produced barium according to the reaction

$$\text{n} + {}^{238}\text{U} \rightarrow {}^{145}\text{Ba} + {}^{94}\text{Kr} \tag{27}$$

The barium and krypton isotopes produced in this fission reaction are very rich in neutrons, and they almost immediately eject several of their excess neutrons. Thus, the net reaction can be summarized as

$$\text{n} + {}^{238}\text{U} \rightarrow \text{fission fragments} + 2\text{n or } 3\text{n} \tag{28}$$

These extra neutrons are crucial for the practical exploitation of fis-

---

* In nuclear physics it is customary to calculate the Q value from the masses of the atoms rather than the masses of the nuclei. This means adding 9 electron masses to each side of Eq. (26), but these electron masses cancel in the equation for Q. For the sake of simplicity, all calculations in this book are done with nuclear masses.

sion, since they can trigger further fission reactions in the uranium sample. This leads to a chain reaction, in which neutrons induce fissions, and the neutrons from these new fissions induce even more fissions, etc. In a nuclear reactor, the chain reaction is brought under control by the deliberate removal of some neutrons by means of absorbers (control rods) immersed in the reactor. In a nuclear bomb, the chain reaction is permitted to grow unchecked, leading to an explosive release of energy.

Fusion reactions, in which two light isotopes merge to form a heavier isotope, are the source of the energy released in the thermonuclear processes in the center of the Sun and also in thermonuclear explosions. In the core of the Sun, fusion reactions proceed through several steps, the net result of which is the fusion of four hydrogen nuclei into one helium nucleus ("hydrogen burning").

Q values are also relevant for the study of **decay reactions**, in which a single particle at rest spontaneously falls apart into two or more particles (we can assume that the initial particle is a rest, because we can always elect to examine the reaction in the rest frame of that particle). Since the initial kinetic energy in a decay reaction is zero, the Q value must be positive if the reaction is to be viable. We can see this from Eq. (24), where $K = 0$ implies $Q = \bar{K} \geq 0$. This gives us a quick way to decide which decay reactions are permitted by energy conservation and which are forbidden. An example of a permitted decay reaction is the beta decay of the neutron, which produces a proton, an electron (or beta particle), and an antineutrino:

$$\mathrm{n} \rightarrow \mathrm{p} + \mathrm{e} + \bar{\nu} \qquad (29)$$

From the masses $m_\mathrm{n} = 939.56\ \mathrm{MeV}/c^2$, $m_\mathrm{p} = 938.27\ \mathrm{MeV}/c^2$, $m_\mathrm{e} = 0.511\ \mathrm{MeV}/c^2$, and $m_\nu = 0$, we find

$$\begin{aligned} Q &= m_\mathrm{n}c^2 - m_\mathrm{p}c^2 - m_\mathrm{e}c^2 - m_\nu c^2 \\ &= 939.56\ \mathrm{MeV} - 938.27\ \mathrm{MeV} - 0.511\ \mathrm{MeV} - 0\ \mathrm{MeV} = 0.78\ \mathrm{MeV} \end{aligned}$$

This positive value of Q confirms that the reaction is permitted.

An example of a forbidden decay reaction is the decay of the electron. Conservation of electric charge demands that if the electron decays, it must decay into a charged particle. But since the electron is the least massive among all the charged particles, the Q value for every conceivable decay into another charged particle is negative, and the decay is forbidden.

5.4 COLLISIONS OF ELEMENTARY PARTICLES; THE CENTER OF MOMENTUM

The most spectacular inelastic collisions occur between high-energy elementary particles, where a large initial kinetic energy is available for conversion into rest mass. The rest mass of each elementary particle is fixed, so the conversion of kinetic energy into rest mass in an inelastic collision between such particles necessarily involves the creation of other elementary particles.

For instance, Fig. 5.3 shows the result of an inelastic collision between two protons. This figure is a photograph of tracks of particles in a bubble chamber filled with liquid hydrogen. When a charged particle passes through the liquid, it produces a fine trail of bubbles, and these reveal its track. The bubble chamber is placed in a magnetic field, which curves the track toward the right or the left, depending on the electric charge of the particle. A high-energy proton produced by the Brookhaven accelerator entered the bubble chamber from below and collided with one of the protons at rest in the nucleus of a hydrogen atom of the liquid. The violent collision destroyed one of the protons but produced sixteen new particles, in part from the rest mass of the destroyed proton and in part by conversion of kinetic energy into mass. The sixteen new particles are seven negative pions, seven positive pions, a positive kaon, and a neutral lambda (invisible, but revealed by its subsequent decay into two charged particles). The production of new particles in this collision can be summarized by the reaction

$$\begin{aligned} \mathrm{p} + \mathrm{p} \rightarrow \mathrm{p} &+ \pi^- + \pi^- + \pi^- + \pi^- + \pi^- + \pi^- + \pi^- \\ &+ \pi^+ + \pi^+ + \pi^+ + \pi^+ + \pi^+ + \pi^+ + \pi^+ + \mathrm{K}^+ + \Lambda^0 \end{aligned} \tag{30}$$

Fig. 5.3 Bubble chamber tracks of particles produced in a proton-proton collision. Because of the magnetic field in the bubble chamber, the tracks of the positively and negatively charged particles produced in this collision curve in opposite directions, positively charged particles toward the right and negatively charged particles toward the left. (Lawrence Berkeley Laboratory)

To initiate the production of the new particles, the incident proton must have not only a kinetic energy large enough to supply the rest-mass energies of the new particles but also large enough to maintain the motion of the center of mass, or, more precisely, the motion of the center of momentum. As in the Newtonian case, the particles that exist after the collision cannot be at rest; they must have an average translational motion such that their total momentum matches the momentum of the incident proton. This means that not all of the kinetic energy of the incident proton is available for creation of new particles--most of the kinetic energy is tied to the motion of the center of momentum, and it is unavailable. Consequently, the threshold energy required for the incident proton is larger than the Q value of the reaction.

We can calculate the threshold energy by using the conservation of energy and momentum. To begin this calculation, we first determine the velocity of the reference frame in which the total momentum of the system of particles is zero. This is called the **center of momentum** reference frame. In Newtonian physics this reference frame is that of the center of mass; but in relativistic physics the center of mass differs from the center of momentum, and only the latter is of importance. We can determine this velocity by recalling the transformation law for the momentum (Eq. 4.37):

$$P'_x = \frac{P_x - VE/c^2}{\sqrt{1 - V^2/c^2}} \tag{31}$$

Here P_x is the sum of particle momenta, and E is the sum of particle energies (either before or after the collision; since momentum and energy are conserved, it makes no difference whether we look at the sums before or after). To make P'_x zero, we must choose

$$\boxed{V = c^2 \frac{P_x}{E}} \tag{32}$$

This is the velocity of the center of momentum. Since both P_x and E are conserved during the collision, the velocity (32) is the same before and after the collision. [Note that in the nonrelativistic limit, the total momentum for a system of, say, two particles is $P_x = m_1 v_{1x} + m_2 v_{2x}$, and the energy E is the sum of the particle rest energies, $E = m_1 c^2 + m_2 c^2$; then Eq. (32) reduces to the familiar formula for the velocity of the center of mass, $V = (m_1 v_{1x} + m_2 v_{2x})/(m_1 + m_2)$.]

When the new particles are produced with the least energy, they will be at rest in the center of momentum frame, that is, they will be at rest relative to each other, and they will travel in company, as a tight bunch (in Fig. 5.3 we see that the new particles created in the collision gradually spread apart; this indicates that they are not at rest relative to one another, and from this we can conclude that the energy of the incident proton was in excess of the threshold energy). For particles at rest in the center of momentum frame, the energy in that frame is simply the sum of rest-mass energies,

$$E' = \Sigma \bar{m}_i c^2 \tag{33}$$

Since P'_x is zero, the inverse transformation law for the energy-momentum four vector gives us

$$E = \frac{E' + cP'_x}{\sqrt{1 - V^2/c^2}} = \frac{\Sigma \bar{m}_i c^2}{\sqrt{1 - (cP_x/E)^2}} \tag{34}$$

If we square both sides of this equation and rearrange, we obtain

$$E^2 - c^2 P_x{}^2 = (\Sigma \bar{m}_i c^2)^2 \tag{35}$$

In the special case of a particle 1 of energy E_1 and momentum p_1 incident on a particle 2 at rest, the initial sum of energies is $\mathrm{E} = E_1 + m_2 c^2$, and the initial sum of momenta is $P_x = p_1$. Therefore, Eq. (35) becomes

$$(E_1 + m_2 c^2)^2 - c^2 p_1{}^2 = (\Sigma \bar{m}_i c^2)^2 \tag{36}$$

or, with $E_1{}^2 - c^2 p_1{}^2 = m_1{}^2 c^4$,

$$2E_1 m_2 c^2 + m_1{}^2 c^4 + m_2{}^2 c^4 = (\Sigma \bar{m}_i c^2)^2 \tag{37}$$

Accordingly, the threshold energy is

$$\boxed{E_1 = \frac{1}{2m_2 c^2}[(\Sigma \bar{m}_i c^2)^2 - m_1{}^2 c^4 - m_2{}^2 c^4]} \tag{38}$$

■ **EXAMPLE 2.** The masses of the particles involved in the reaction (30) are 0.938 GeV/c^2 for the proton, 0.140 GeV/c^2 for each pion, 0.494 GeV/c^2 for the kaon, and 1.115 GeV/c^2 for the lambda. What is the threshold energy for the reaction? What is the threshold kinetic energy?

SOLUTION. The final particles are one proton, fourteen pions, one kaon, and one lambda. The sum of final rest-mass energies is

$$\Sigma \bar{m}_i c^2 = 0.938 \text{ GeV} + 14 \times 0.140 \text{ GeV} + 0.494 \text{ GeV} + 1.115 \text{ GeV} = 4.507 \text{ GeV}$$

With $m_1 = m_2 = 0.938$ GeV/c^2, Eq. (38) then gives a threshold energy

$$E_1 = \frac{1}{2 \times 0.938 \text{ GeV}} [(4.507 \text{ GeV})^2 - 2 \times (0.938 \text{ GeV})^2] = 9.890 \text{ GeV}$$

The threshold kinetic energy is

$$K_1 = E_1 - m_1c^2 = 9.890 \text{ GeV} - 0.938 \text{ GeV} = 8.952 \text{ GeV}$$

This example indicates that the creation of new particles by the impact of a projectile on a stationary target is very inefficient. In the reaction (9) we create 2.631 GeV of new rest-mass energy (the difference between the initial rest mass and the final rest mass), but we need a projectile of kinetic energy 8.952 GeV to accomplish this. Most of this kinetic energy is locked up in the motion of the center of momentum; it is "useless" energy.

In their quest for more and more new particles of larger and larger rest masses, experimental physicists built larger and larger accelerators. The largest accelerators built so far, at CERN and at Fermilab, supply protons and antiprotons of energy 1 TeV = 10^3 GeV. The impact of a proton of this energy on a stationary target proton permits the creation of particles with a sum of rest masses no larger than the limit set by Eq. (38), that is, the value of $\Sigma\, \bar{m}_i c^2$ is at most

$$\Sigma\, \bar{m}_i c^2 = \sqrt{2m_2c^2(E_1 + m_2c^2)} \simeq \sqrt{2m_2c^2E_1} \tag{39}$$

Fig. 5.4 At the Fermilab accelerator, 120-GeV protons from the Main Injector (oval at lower left) are first used to produce antiprotons, by collision with a target. Bunches of protons and antiprotons traveling in opposite directions are then accelerated and concentrated in the Main Injector, and they are injected into the Tevatron (circle at center). There they are accelerated to their final energy of 1 TeV, and they are brought into head-on collision. (Fermilab)

$$= \sqrt{2 \times 0.938 \text{ GeV} \times 10^3 \text{ GeV}} = 43 \text{ GeV}$$

Particles of higher rest-mass energy than that cannot be produced.

To get around the limitations imposed by the motion of the center of momentum, physicists now operate the large accelerators as **colliders**. They accelerate particles and antiparticles--for instance, protons and antiprotons--and bring these together in head-on collisions (see Fig. 5.4). The momenta of the colliding protons are then of equal magnitudes and opposite directions, and the net momentum is zero; thus, the center of momentum frame is at rest in the laboratory, and all of the energy of both protons is available for conversion into rest mass. At Fermilab, this permits the creation of particles with a sum of rest masses as large as 1 TeV + 1 TeV = 2 TeV = 2×10^3 GeV. The heaviest particles discovered so far, the W and Z bosons, have rest masses of about 80 GeV/c^2, and they were produced in such collider experiments.

5.5 RECOIL IN PHOTON EMISSION; THE MÖSSBAUER EFFECT

When an atom or a nucleus emits a photon, the atom or nucleus must provide not only the energy of the photon but also its momentum. Consequently, the atom or nucleus suffers a recoil quite analogous to the recoil of gun when it fires a bullet. For photons in the visible range emitted by atoms, the momentum is fairly small, and the recoil is unimportant. But for the high-energy gamma rays emitted by nuclei, the recoil can be significant.

The main consequence of such a recoil is that the photon is emitted with a reduced energy, since the nucleus acquires some of the energy that would be carried off by the photon if there were no recoil. We can easily calculate the recoil energy from the conservation of energy and momentum. Suppose that the rest mass of the nucleus before the emission of the photon is m and that the rest mass and the energy after the emission are $\bar{m}$ and $\bar{E}$, respectively. Suppose that the energy of the emitted photon is W. Then energy conservation tells us that

$$\bar{E} = mc^2 - W \tag{40}$$

Suppose that the momentum of the nucleus after the emission is $\bar{p}$. The momentum of the photon is W/c, and momentum conservation tells us that

$$\bar{p} = - W/c \tag{41}$$

We can substitute the energy given by Eq. (40) and the momentum given by Eq. (41) into the identity $\bar{E}^2 = c^2\bar{p} + \bar{m}^2c^4$, with the result

$$m^2c^4 - 2Wmc^2 + W^2 = W^2 + \bar{m}^2c^4 \tag{42}$$

from which we find the photon energy

$$W = \frac{1}{2mc^2}(m^2c^4 - \bar{m}^2c^4) \tag{43}$$

We can give this equation a more instructive form by expressing the right side in terms of the energy W_0 that the photon would have if there were no recoil. This energy is $W_0 = mc^2 - \bar{m}c^2$, from which

$$\bar{m}^2c^4 = W_0^2 - 2mc^2W_0 + m^2c^4 \tag{44}$$

If we insert this into Eq. (43), we find

$$W = W_0 - \frac{W_0^2}{2mc^2} = \left(1 - \frac{W_0}{2mc^2}\right)W_0 \tag{45}$$

Thus, the recoil reduces the photon energy by a factor of $(1 - W_0/2mc^2)$ compared with the energy it would have if there were no recoil.

Numerically, this factor is very close to 1 if $mc^2 >> W_0$. For instance, for photons of visible light emitted by an atom, the energy W_0 is a few electron-volts, and the mass of the atom is typically 100 GeV/c^2; thus, the reduction of the photon energy by recoil is typically 1 part in 10^{11}. This is insignificant, because the photons emitted by atoms have inherent uncertainties in their energy ("line widths") much larger than 1 part in 10^{11}.

But for gamma-ray photons emitted by a nucleus, where the energy W_0 is, say, 10 keV, the reduction of the photon energy is 1 part in 10^7. This is significant, since the photons emitted by some nuclei have line widths that are much smaller. For instance, the 14.4-keV gamma ray emitted by an excited state of the isotope ^{57}Fe has a line width of only 6 parts in 10^{13}. In view of this extremely low uncertainty in the energy of the photon, the reduction of energy by recoil becomes detectable in resonance experiments. For this purpose, the photons emitted by a sample of ^{57}Fe nuclei in the excited state are aimed at a second sample of ^{57}Fe nuclei in the ground state. The second sample would then resonantly absorb the photons if their energy were unchanged or if it were changed by no more than the line width. But the reduction of energy by recoil exceeds these limitations, and absorption becomes impossible (note that there is a recoil at the emitter and a second recoil at the absorber!).

When the nuclei emitting the gamma rays are in a crystalline solid, it is possible for the recoil momentum to be distributed over the entire crystal, instead of being concentrated on the single nucleus that emits the photon. Since the mass of the crystal is much larger than the mass of a nucleus, the reduction of the energy of the photon becomes nearly zero [if m in Eq. (45) is, say, 10^{10} times the mass of an atom, then $W_0/m \simeq 10^{-17} \simeq 0$]. Under these circumstances, the emission of the photon is practically recoilless. This kind of emission was discovered by R. Mössbauer in 1958, and it is called the **Mössbauer effect.** The 14.4-keV recoilless photons emitted by ^{57}Fe provide us with an extremely precise frequency standard, which has been exploited in high-precision tests of Special Relativity and of General Relativity, such as the tests of time dilation described in Chapters 3 and 7.

5.6 RELATIVISTIC ROCKET MOTION

In this section we will apply relativistic mechanics to a simple case of motion of a macroscopic body, a self-propelled rocket that attains a relativistic speed. Although this case does not yet have practical applications, it is instructive in that it tells us something about the limitations of interstellar travel.

Rocket propulsion operates by recoil. High-speed gas is ejected from the tail end of the rocket, and the recoil pushes the rocket forward. The most efficient propulsion system for a rocket relies on the ejection of light from the tail of the rocket, instead of the ejection of gas. For a given amount of energy, this produces the maximum recoil. We can see this from the relation between energy and momentum of a mass m of ejecta,

$$c^2p^2 = E^2 - m^2c^4 \tag{46}$$

For a given energy E, the momentum p (and the opposite recoil momentum $-p$) is maximum if $m = 0$.

For maximum efficiency we need a fuel that can be converted totally into light or, more generally, into radiant energy. In principle, the best choice is an equal mixture of matter and antimatter, which produces gamma rays. Unfortunately, antimatter is not available in bulk quantities (only minuscule amounts of antimatter have been produced in accelerators). And furthermore, the annihilation of matter and antimatter produces pairs of gamma rays that travel in opposite directions, and it is not clear how to construct a practical device that reflects and guides the gamma rays so that they all emerge in a parallel beam from the tail of the rocket. But let us ignore these practical considerations and examine what limitations the conservation of energy and momentum imposes on the final speed attainable by a rocket with "light propulsion."

Suppose that the initial mass of the rocket (including the engine and the matter-antimatter fuel) is m_i and that the final mass at burn-

Fig. 5.5 At burnout, the emitted radiation has a total energy $E_{\rm rad}$ and a momentum $E_{\rm rad}/c$ toward the left. The rocket has a momentum toward the right.

out is $m_{\rm f}$. The initial speed is 0 and the final speed is $v_{\rm f}$. The initial energy is simply the rest-mass energy, and the final energy is the sum of the energy of the rocket plus the energy $E_{\rm rad}$ of all of the gamma radiation emitted (see Fig. 5.5). Conservation of energy then demands

$$m_{\rm i}c^2 = \frac{m_{\rm f}c^2}{\sqrt{1 - v_{\rm f}^2/c^2}} + E_{\rm rad} \qquad (47)$$

Likewise, conservation of momentum demands

$$0 = \frac{m_{\rm f}v_{\rm f}}{\sqrt{1 - v_{\rm f}^2/c^2}} - \frac{E_{\rm rad}}{c} \qquad (48)$$

By eliminating $E_{\rm rad}$ between these equations, we find

$$m_{\rm i}c = \frac{m_{\rm f}c}{\sqrt{1 - v_{\rm f}^2/c^2}} + \frac{m_{\rm f}v_{\rm f}}{\sqrt{1 - v_{\rm f}^2/c^2}} \qquad (49)$$

If we square both sides of this equation, we obtain a quadratic equation for the final speed $v_{\rm f}$,

$$m_{\rm i}^2\,(1 - v_{\rm f}^2/c^2) = m_{\rm f}^2\,(1 + v_{\rm f}/c)^2$$

or

$$m_{\rm i}^2\,(1 - v_{\rm f}/c) = m_{\rm f}^2\,(1 + v_{\rm f}/c) \qquad (50)$$

We can solve this equation for the final speed:

$$v_{\rm f} = c \times \frac{(m_{\rm i}/m_{\rm f})^2 - 1}{(m_{\rm i}/m_{\rm f})^2 + 1} \qquad (51)$$

This is the relativistic "rocket equation"; it tells us the final speed as a function of the ratio of initial mass to final mass. Alternatively, we can solve Eq. (50) for the mass ratio:

$$\frac{m_i}{m_f} = \sqrt{\frac{1 + v_f/c}{1 - v_f/c}} \tag{52}$$

This tells us the mass ratio as a function of the final speed.

■ **EXAMPLE 3.** Suppose the mass of a rocket is 100 tons at burnout. What initial mass of matter-antimatter fuel is required for this rocket to attain a final speed of $0.9c$? a final speed of $0.99c$?

SOLUTION. For $v_f = 0.9c$, Eq. (52) gives

$$\frac{m_i}{m_f} = \sqrt{\frac{1 + 0.9}{1 - 0.9}} = 4.4$$

Hence $m_i = 4.4 \times 100$ tons $= 440$ tons, and the required mass of fuel is 440 tons - 100 tons = 340 tons.

Likewise, for $v_f = 0.99c$,

$$\frac{m_i}{m_f} = \sqrt{\frac{1 + 0.99}{1 - 0.99}} = 14.1$$

Hence $m_i = 1410$ tons, and the required mass of fuel is 1310 tons.

Although the amounts of fuel calculated in Example 3 seem modest, they are gigantic compared with the minuscule amount of antimatter that has been produced in accelerators (the Fermilab accelerator produces bunches of antiprotons of about 10^{-15} kg). Furthermore, if we want to attain the final relativistic speed in a reasonable amount of time, say 10 years, the engine must cope with a gigantic amount of power. The "burning" of 340 tons of matter-antimatter in, say, 10 years implies that the engine must cope with a power of about 10^{14} watts, which is about a hundred times the total power used in the entire United States. And the engine coping with this power must be small--it must itself have a mass of no more than a few tons. Because of these limitations, most physicists believe that interstellar travel with high-speed, relativistic self-propelled rockets is not feasible.

A scheme for overcoming the fuel limitations on high-speed interstellar travel was examined by Bussard [1], who proposed exploiting the interstellar hydrogen gas by means of a "ramjet." Such a spacecraft would have a large funnel at its front to collect the tenuous interstellar gas (average density 1 hydrogen atom per cubic centimeter), and it would use this gas as fuel in a fusion reactor. The spent fuel would then be ejected at the rear of the ramjet with a higher momentum than the momentum with which it entered, and this would provide propulsion. The collecting funnel would have to be more than 1 km across for a spacecraft of 100 tons. And the funnel would probably have to consist of suitably shaped magnetic fields, since a sheet of ordinary material would have to be prohibitively thick to resist penetration by hydrogen atoms incident on the funnel with a relative speed comparable with the speed of light. Although the broad features of the ramjet are consistent with the laws of conservation of energy and momentum, the details require technologies that are far beyond our current knowledge.

REFERENCES

1. I. S. Shklovskii and C. Sagan, *Intelligent Life in the Universe* (Holden-Day, New York, 1966), p. 445.

PROBLEMS

1. Show that according to Eq. (8), $\bar{p}_1 > 0$ if $m_1 > m_2$ and $\bar{p}_1 < 0$ if $m_1 < m_2$.

2. Equations (8) and (9) give the momenta after an elastic collision between an incident projectile and an initially stationary target.
(a) Verify that for a collision of particles of equal masses, the target particle acquires all of the kinetic energy, and the projecticle remains at rest. (This is agreement with the well-known Newtonian result for elastic collisions between particles of equal masses.)
(b) Consider a projectile of zero mass (for instance, a photon). Show that such a projectile always recoils, reversing its motion. Show that if the incident projectile has a momentum much larger than m_2c, its recoil momentum is $m_2c/2$. This means that almost all of the momentum of the incident projectile is transferred to the target particle.
(c) Consider a photon of energy 1 GeV incident on an initially stationary electron. If the collision is on a straight line and elastic, what is the photon energy after the collision? (Compare Problem 9).

3. A particle of mass m and initial kinetic energy K_1 collides elastically with an identical particle initially at rest. Suppose that after this collision, the two particles emerge symmetrically, with equal energies and at equal angles with respect to the initial direction of motion, one to the right, one to the left.
(a) Calculate the angle θ at which the particles emerge; express your answer in terms of K_1 and m.
(b) What is the angle θ if $K_1 = 10\ mc^2$? if $K_1 = mc^2$? if $K_1 = 0.10\ mc^2$?
(c) Show that if $K_1 \ll mc^2$ the angle is $\theta = 45^0$, so the angle between the particles is 90^0.

4. Show that if $m_1 = 0$, the result (8) for an elastic collision on a straight line

agrees with the Compton result (21) at $\theta = 180^0$.

5. A photon of initial wavelength 5.0×10^{-11} m suffers a Compton collision with a free electron and is deflected by an angle of 90^0. What is the final wavelength of the photon? What is the energy lost by the photon? What is the energy gained by the electron?

6. Photons of energy 40 keV are incident on a target containing (almost) free electrons. What is the maximum energy that an electron can acquire by recoil in a collision with such a photon?

7. A photon of initial wavelength 1.0×10^{-11} m collides with two electrons in succession. The photon is deflected by 45^0 in the first collision and by 30^0 in the second collision. What is the wavelength of the photon after the first collision? after the second?

8. A photon of initial energy 2 0 MeV collides with a *proton* and is deflected by 30^0. What is the final energy of the photon?

9. A photon of energy much larger than 0.511 MeV collides with a free electron at rest and recoils backward, emerging at $\theta = 180^0$. Show that in this case, the recoiling photon will have an energy of about 0.205 MeV and that the recoiling electron will acquire almost all of the large energy of the photon.

10. Show that the momentum and the kinetic energy of the recoiling electron in the Compton effect are

$$\bar{p} = 2m_e c \frac{(h\nu + m_e c^2)h\nu \cos\phi}{(h\nu + m_e c^2)^2 - (h\nu \cos\phi)^2}$$

$$\bar{K} = m_e c^2 \frac{2(h\nu \cos\phi)^2}{(h\nu + m_e c^2)^2 - (h\nu \cos\phi)^2}$$

Here ϕ is the angle of the motion of the electron relative to the incident photon. Verify that $\bar{K}$ is maximum for $\phi = 0$ and minimum for $\phi = 90^0$.

11. In **inverse Compton scattering** a moving electron collides elastically with a photon, and both the electron and the photon recoil. Consider the simple case of a head-on collision between an electron of momentum p_1 moving toward the left and a photon of momentum p_2 moving toward the right on the same straight line of motion.

(a) Find the momenta $\bar{p}_1$ and $\bar{p}_2$ of the electron and the photon after the collision. (*Hint*: The photon reverses its direction of motion during the collision).

(b) Show that if the electron is initially extremely relativistic ($E_1 \gg m_e c^2$), then $\bar{p}_2 \simeq p_1$ and $\bar{p}_1 \simeq p_2$. This means that the electron and the photon merely exchange momenta during the collision.

12. An electron and an antielectron of nearly zero kinetic energy collide and annihilate, producing two photons. What is the energy of each photon? What is the momentum of each? If one photon moves eastward, what is the direction of motion of the other?

13. Consider the emission of a gamma ray by an electron and consider the annihilation of an electron and an antielectron into a single gamma ray, according to the reactions

$$\mathrm{e} \rightarrow \mathrm{e} + \gamma$$

$$\mathrm{e} + \bar{\mathrm{e}} \rightarrow \gamma$$

Show that one of these reactions is forbidden by energy conservation and the other by momentum conservation.

14. Calculate the Q value for the decay of helium into deuterium, according to the reaction

$$^4\mathrm{He} \rightarrow {}^2\mathrm{H} + {}^2\mathrm{H}$$

The mass of the deuterium, or ^{2}H, nucleus is 1875.6 MeV/c^2; the mass of ^{4}He is given in Section 5.3. Can this decay proceed?

15. What are the maximum kinetic energy and the maximum speed of the electrons emitted in the beta decay of a neutron, initially at rest, according to the reaction (29)? What are the kinetic energy and the speed of the recoiling proton in this case? (*Hint*: To give the electron maximum energy, the neutrino must emerge with zero energy.)

16. Calculate the Q values for the reactions

$$^6\mathrm{Be} \rightarrow {}^5\mathrm{Li} + \mathrm{p}$$

$$^5\mathrm{Li} \rightarrow {}^4\mathrm{He} + \mathrm{p}$$

The masses of the ^{6}Be and ^{5}Li nuclei are 5605.3 MeV/c^2 and 4667.6 MeV/c^2, respectively; the masses of ^{4}He and p are given in Section 5.3. What can you conclude about the stability of ^{6}Be and of ^{5}Li?

17. The isotope ^{13}N was discovered by Curie and Joliot by the bombardment of boron with alpha particles, leading to the reaction

$$^4\mathrm{He} + {}^{10}\mathrm{B} \rightarrow {}^{13}\mathrm{N} + \mathrm{n}$$

The masses of the ^{10}B and ^{13}N nuclei are 9324.4 MeV/c^2 and 12,111.2 MeV/c^2, respectively; the masses of the alpha particle, or ^{4}He, and of the neutron are given in Section 5.3. What is the Q value of this reaction? Can it proceed with alpha particles of nearly zero energy?

18. The old Cosmotron at the Brookhaven Laboratory produced protons of kinetic energy 2.3 GeV, which were used to bombard target protons at rest. What is the velocity of the center of momentum of the system that consists of such a moving proton and a target proton? Compare this velocity with the velocity of the center of mass calculated according to Newtonian physics.

19. A gamma ray of energy E_1 is traveling along the x axis. A second gamma ray of energy E_2 is traveling in the x-y plane at angle of θ with respect to the first. What is the velocity of the center of momentum of this system of two gamma rays? Under what conditions is this velocity zero?

20. A π^0 meson of kinetic energy 200 MeV decays into two gamma rays. One gamma ray is emitted in the direction of motion of the meson, and one is emitted in the opposite direction. What is the Q value for the reaction $\pi^0 \rightarrow \gamma + \gamma$? What is the energy of each gamma ray in the rest frame of the π_0? What is the energy of each gamma ray in the laboratory frame? The mass of the π^0 is 135 MeV/c^2.

21. In Example 2 we found that to initiate the reaction (30) by the impact of a high-energy proton on a stationary target proton, we need a kinetic energy 8.952 GeV. If, instead, we initiate the reaction (30) by means of a head-on collision of two protons, what is the minimum kinetic energy of each proton?

22. The Superconducting Supercollider (SSC), whose construction was begun and later abandoned near Dallas, Texas, was supposed to accelerate protons to energies of 20 TeV. If such a proton is incident on a stationary target proton, what is the maximum possible value of the sum of rest-mass energies $\Sigma\, \bar{m}_i c^2$ of the final particles? What is this maximum value for a head-on collision between two protons of 20 TeV each?

23. From Eq. (38), show that for an endoergic reaction the kinetic energy K_1 at threshold is always larger than $|Q|$.

24. Show that if the particle energies are nonrelativistic, then Eq. (38) leads to a threshold kinetic energy of

$$|Q| \times \frac{\Sigma\, \bar{m}_i}{\Sigma\, \bar{m}_i - m_1}$$

25. What is the threshold kinetic energy for the reaction (25)?

26. The W bosons and antibosons were first produced at the CERN Super Proton Synchrotron, where protons and antiprotons were made to collide head-on with an energy of 270 GeV each. One of the reactions triggered by such collisions was

$$\mathrm{p} + \bar{\mathrm{p}} \rightarrow \mathrm{W}^+ + \mathrm{W}^-$$

which converts a proton-antiproton pair into a W boson-antiboson pair. These W bosons are extremely heavy particles, with a mass of 80 GeV/c^2. Calculate the kinetic energy and the speed of the bosons produced in the reaction.

27. Consider the production of a W boson-antiboson pair by the collision of a proton and antiproton, according to the reaction described in the preceding problem. If instead of using a head-on collision, we want to use a high-energy antiproton incident on a proton at rest, what antiproton energy do we need?

28. A beam of π^- mesons is incident on a target containing protons at rest. Find the threshold kinetic energy of the π^- meson to initiate the reaction

$$\pi^- + p \rightarrow \Lambda^0 + K^0$$

The masses of the π^-, Λ^0, and K^0 are 0.140 MeV/c^2, 1.115 GeV/c^2, and 0.498 GeV/c^2, respectively.

29. A beam of K^- mesons is used to bombard a target of protons at rest. For each of the following reactions, find the threshold kinetic energy for the K^- meson to initiate the reaction:

(a) $$K^- + p \rightarrow \Xi^0 + K^0$$

(b) $$K^- + p \rightarrow \Lambda^0 + \pi^+ + \pi^-$$

The masses of the K^-, K^0, Ξ^0, Λ^0, and $\pi^\pm$ are 0.494 GeV/c^2, 0.498 GeV/c^2, 1.315 GeV/c^2, 1.115 GeV/c^2, and 0.140 GeV/c^2, respectively.

30. We want to produce a τ lepton and antilepton pair by the collision of electrons and antielectrons according to the reaction

$$e + \bar{e} \rightarrow \tau + \bar{\tau}$$

The τ lepton has a mass of 1.784 GeV/c^2. What is the threshold kinetic energy required to initiate the reaction (a) if we use electron and antielectrons beams in a head-on collision? (b) if we use a beam of antielectrons to bombard a target containing stationary electrons?

31. (a) A particle of mass m_1 at rest decays into several particles of masses $\bar{m}_1$, $\bar{m}_2$, . . . Show that the maximum kinetic energy that an emerging particle k of mass $\bar{m}_k$ can acquire in this decay is

$$\left(1 - \frac{\bar{m}_k}{m_1} - \frac{Q}{2m_1c^2}\right) Q$$

(*Hint*: The particle k acquires maximum energy when all the other particles created in the decay remain together and recoil in the direction opposite to that of particle k.)

(b) Consider the decay reaction

$$K^+ \rightarrow \pi^+ + \pi^0 + \pi^0$$

The mass of the K^+ is 0.494 GeV/c^2, and the masses of the π^+ and the π^0 are 0.140 GeV/c^2 and 0.135 GeV/c^2, respectively. Find the maximum kinetic energy of the π^+ emerging from such a decay.
(c) Find the maximum kinetic energy of the π^0 emerging from such a decay.

32. The preceding problem gives a general formula for the maximum kinetic energy of a particle emerging in a decay. Show that this formula yields the result (45) when the "decay" is the emission of a photon.

33. The ^{7}Be nucleus, of mass 6534 MeV/c^2, is observed to emit a gamma ray of energy 0.478 MeV. What are the recoil energy and speed of the nucleus? How much larger would the gamma ray energy have been if the nucleus had not recoiled?

34. A nucleus, initially at rest, absorbs a gamma ray of energy W. Taking into account the recoil of the nucleus, express the increase of rest mass of the nucleus in terms of the gamma-ray energy W and the final mass m of the nucleus.

35. Consider a rocket propelled by means of a chemical fuel. The hot gas produced by the combustion of this fuel is ejected from the tail of the rocket at a speed u_{ex} relative to the rocket. The initial mass of the rocket, including the fuel, is m_i, and the final mass, after combustion of all the fuel, is m_f. Show that the final velocity v_f attained by this rocket is given by

$$\frac{m_i}{m_f} = \left[\frac{1 + v_f/c}{1 - v_f/c}\right]^{c/2u_{ex}}$$

Show that if $u_{ex} = c$, this reduces to the formula given in Section 5.5, and show that if $u_{ex}/c \ll 1$, it reduces to the familiar rocket formula of Newtonian physics,

$$v_f = u_{ex} \ln \frac{m_i}{m_f}$$

[*Hint*: In the instantaneous rest frame of the rocket at some given time, the equations for the conservation of energy and momentum during the ejection of a small mass dm of exhaust are

$$-c^2 dm = \frac{c^2 dm}{\sqrt{1 - v_{ex}^2/c^2}} \quad \text{and} \quad m dv' = \frac{v_{ex} dm}{\sqrt{1 - v_{ex}^2/c^2}}$$

Furthermore, according to the combination law for velocities, the increments dv in the laboratory frame and dv' in the instantaneous rest frame are related by $v + dv = (v + dv')/(1 + v dv'/c^2)$.]

36. Instead of using rocket propulsion for a spacecraft, it is more efficient to propel the spacecraft by the push of a laser beam aimed at a "push plate" mounted at its tail end. The fuel and the engine that produces the laser beam can then remain on Earth, and the only mass that needs to be propelled is the payload and the push plate.

(a) Suppose that an amount of light energy dE_{rad} (reckoned in the reference frame of the Earth) is incident on the push plate and is totally reflected by the push plate. Show that if the instantaneous speed of the spacecraft is v, then the equation of motion is

$$\frac{d}{dt}\frac{mv}{\sqrt{1-v^2/c^2}} = \left(1 + \frac{1-v/c}{1+v/c}\right)\frac{dE_{\text{rad}}}{cdt}$$

(*Hint*: For the reflected light, take into account the Doppler shift for incidence *and* the Doppler shift for re-emission by the push plate).

(b) Integrate the equation of motion (with a constant mass m!) and obtain the equation for the final speed:

$$v_{\text{f}} = c \times \frac{(2E_{\text{rad}}/mc^2 + 1)^2 - 1}{(2E_{\text{rad}}/mc^2 + 1)^2 + 1}$$

(c) What is the value of E_{rad} required to accelerate a spacecraft of 100 tons to a speed of 0.9c? to a speed of 0.99c? Compare E_{rad} with the final kinetic energy of the spacecraft. Also compare with the results of Example 3.

6

Electromagnetism and Relativity

I HAVE LITTLE PATIENCE WITH SCIENTISTS WHO TAKE A BOARD OF WOOD, LOOK FOR ITS THINNEST PART AND DRILL A GREAT NUMBER OF HOLES WHERE DRILLING IS EASY.

Albert Einstein to Phillip Frank

Electromagnetic theory played a crucial role in the genesis of relativity. Einstein discovered Special Relativity by thinking about the transformation of Maxwell's equations from one inertial reference frame to another. He called his first publication on relativity "On the Electrodynamics of Moving Bodies," and after deriving the Lorentz transformation in this paper, he immediately applied it to the transformation of the electric and magnetic fields and to the transformation of Maxwell's equations.

In this chapter we will not imitate Einstein's approach to electromagnetism, because the general transformation formulas for the complete set of Maxwell equations are somewhat complicated. Instead, we will adopt a simpler approach, proceeding by way of special examples in which it is easy to find the relativistic transformation laws for the electric and magnetic fields. We will see that under these transformation laws, electric and magnetic fields transform into one another, just as space and time transform into one another under the Lorentz transformation. What is a purely electric field (or a purely magnetic field) in one inertial reference frame becomes a combination of electric and magnetic fields in another inertial reference frame.

We will also see that relativity permits us to deduce the electric and magnetic fields of a uniformly moving charge and of an accelerated charge from Coulomb's Law. This means that relativity provides a tight link between the four Maxwell equations--Gauss' Law for electric fields, Gauss' Law for magnetic fields, Faraday's Law, and the Maxwell-Ampère Induction Law*. With some reasonable assumptions, the four Maxwell equations can all be deduced from the first Maxwell equation, that is, from Gauss' Law for the electric fields, or, equivalently, from Coulomb's Law.

6.1 THE ELECTRIC FIELD OF A MOVING CHARGE

As a first application of relativity in electromagnetism, we investigate the electric field of a point charge q moving with uniform velocity. We can deduce this electric field by beginning with the known electric field in the rest frame of the point charge. In this reference frame, the charge is at rest, and the electric field is simply the familiar inverse-square field given by Coulomb's Law. This electric field is directed radially, and it has a magnitude (in SI units)

$$E = \frac{1}{4\pi\epsilon_0}\frac{q}{r^2} \tag{1}$$

We can represent this electric field by radial field lines, uniformly distributed over all angles, or, more precisely, over all solid angles (see Fig. 6.1).

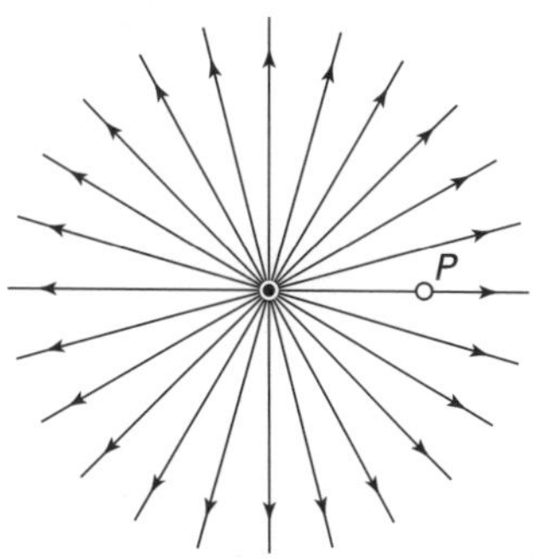

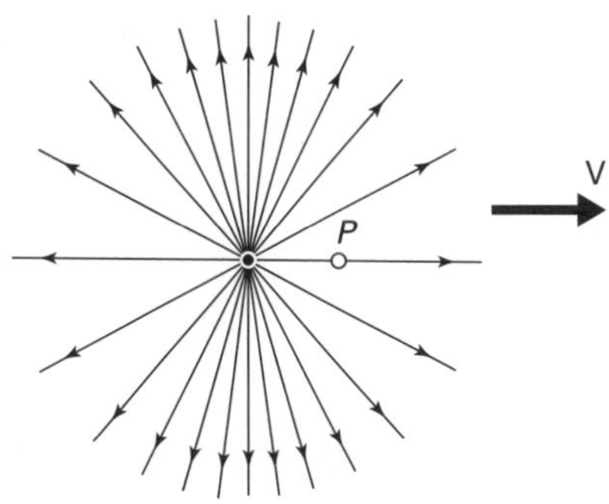

Fig. 6.1 (Left) Electric field lines of a point charge at rest. This diagram shows the field lines in only two dimensions, that is, it shows the field lines in a plane containing the point charge. The field lines are uniformly distributed over all directions in this plane. In three dimensions, the field lines are uniformly distributed over all directions of space. A point P at a given distance to the right of the charge is marked for comparison with the next diagram.

Fig. 6.2 (Right) Electric field lines of a point charge in uniform motion toward the right. The pattern is contracted along the direction of motion. The distance to the point P marked in Fig. 6.1 is also contracted.

---

* Expressed in their integrated form, these four Maxwell equations are $\int \mathbf{E} \cdot d\mathbf{S} = Q/\epsilon_0$, $\int \mathbf{B} \cdot d\mathbf{S} = 0$, $\int \mathbf{E} \cdot d\mathbf{l} = -\,d\Phi_B/dt$, and $\int \mathbf{B} \cdot d\mathbf{l} = \mu_0 I + \mu_0\epsilon_0 d\Phi_E/dt$ (in SI units). However, we will have no occasion to use these equations explicitly in this chapter.

Now imagine that we observe this charge and its electric field from a reference frame that, relative to the charge, has a uniform velocity V toward the left. Relative to this reference frame, the charge has a uniform velocity V toward the right. If we assume that the field lines in Fig. 6.1 are rigidly attached to the charge and participate in its motion, then the pattern of field lines will suffer a length contraction along the direction of motion. This means that the pattern of field lines will be compressed along the x direction by the length-contraction factor $\sqrt{1 - V^2/c^2}$, whereas it will remain uncompressed along the y and z directions. Obviously, this changes the spherically symmetric radial pattern into an asymmetric pattern (see Fig. 6.2), with an enhanced density of field lines perpendicular to the direction of motion and an unchanged density of field lines parallel to the direction of motion. Accordingly, at any location in the field, the component of the electric field perpendicular to the direction of motion is increased by a factor $1/\sqrt{1 - V^2/c^2}$, and the component parallel to the direction of motion remains unchanged. With the usual notation $\gamma = 1/\sqrt{1 - V^2/c^2}$, we have

$$\boxed{\mathbf{E}'_{\parallel} = \mathbf{E}_{\parallel}} \tag{2}$$

$$\boxed{\mathbf{E}'_{\perp} = \gamma \mathbf{E}_{\perp}} \tag{3}$$

Here the subscript $\parallel$ indicates the part of the vector parallel to the velocity, and the subscript $\perp$ indicates the part of the vector perpendicular to the velocity.

Note that in Fig. 6.2, it seems that the strength of the electric field in front of the charge is decreased; however, a given point P at a location in front of the charge will be nearer to the charge in the contracted diagram than in the uncontracted diagram, and the parallel component of the electric field at the given point P is the same in both diagrams.

In the above discussion, we assumed that the field lines behaved as though they were rigidly attached to the charge, "like the quills of a porcupine." This permitted us to apply the relativistic length contraction, as for any rigid body. But this assumption of rigidity is not self-evident, since the field lines are not physical objects but merely mathematical artifacts for the graphical representation of electric fields. To justify our assumption about the contraction of the pattern of field lines for the moving charge, consider a point charge q surrounded by a concentric spherical shell of rigid dielectric over which a charge $-q$ is uniformly distributed (see Fig. 6.3). The field lines that start on the point charge q then end on this shell. If we observe the point charge and the shell from a reference frame with respect to which the point charge and the shell are in motion, then the length contraction is necessarily valid for the shell (see Fig. 6.4). From this we can conclude that the charge distribution on the shell, and therefore also the *endpoints* of the field lines, must be arranged in accord with the length contraction. If the field lines are straight lines, their pattern must then be the contracted pattern displayed in Fig. 6.2. To establish that the field lines are indeed straight and that the intermediate points on each field line are arranged in accord with the length contraction, consider a point charge surrounded by many concentric spherical dielectric shells of different radii, each endowed with some fraction of the charge $-q$. The shorter field lines that end on the smaller of these shells must then have endpoints arranged in accord with the length contraction, and since the longer field lines that end on the outermost shell cannot cross the shorter field lines, they must intersect each shell at a location that is in accord with the length contraction.

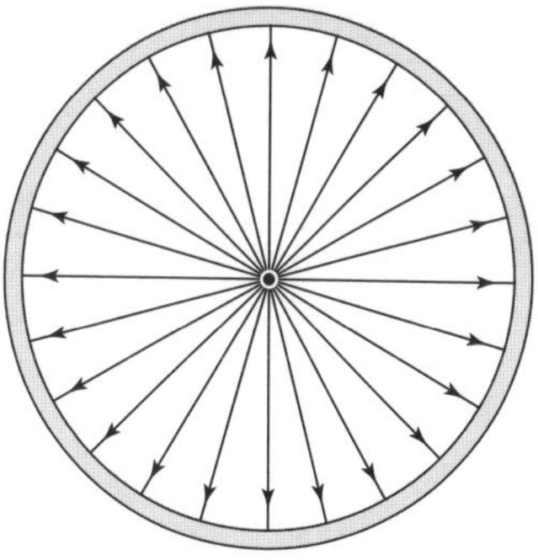

Fig. 6.3 (Left) A point charge q surrounded by a rigid spherical dielectric shell of charge $-q$. The point charge and the shell are at rest.

Fig. 6.4 (Right) The point charge and the shell are in motion toward the right. The shell is now contracted along the direction of motion.

The transformation equations (2) and (3) for the perpendicular and the parallel components of the electric field are also valid for electric fields generated by any arbitrary distribution of charges that are at rest relative to each other. We can establish this by regarding the arbitrary distribution of charges as a collection of point charges. Since Eqs. (2) and (3) are valid for the electric fields of the individual point charges, these equations must also be valid for the net electric field constructed by superposition of the individual electric fields.

Alternatively, we can establish Eqs. (2) and (3) by regarding the electric field in a small region as a uniform electric field generated by a small parallel-plate capacitor. If such a capacitor moves relative to us in a direction perpendicular to that of its plates (so its electric field is parallel to the motion; see Fig. 6.5a), then the electric charge distribution on its plates will remain unchanged, and so will the electric field, in agreement with Eq. (2). If such a capacitor moves relative to us in a direction parallel to that of its plates (so its electric field is perpendicular to the motion; see Fig. 6.5b), then the electric charge distribution on its plates will suffer a length contraction, which increases the charge density, and consequently the electric field, by a factor of $1/\sqrt{1 - V^2/c^2}$, in agreement with Eq. (3).

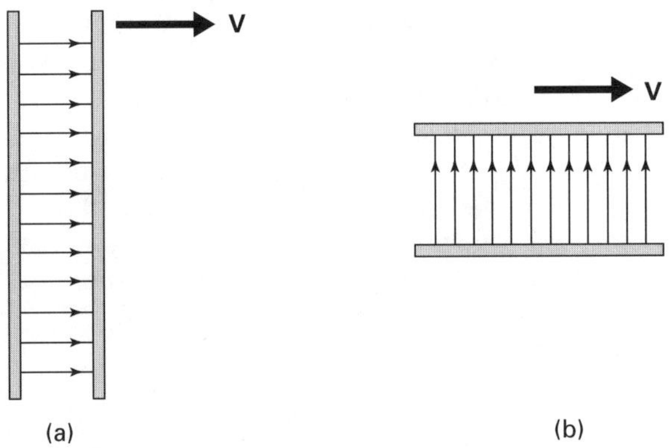

Fig. 6.5 (a) A small capacitor in motion in a direction perpendicular to its plates. The distance between the plates is contracted, but this does not affect either the charge distribution on the plates or the electric field. (b) A small capacitor in motion in a direction parallel to its plates. The areas of the plates are now contracted, the charge densities on the plates are increased, and so is the strength of the electric field.

Note that all these arguments hinge on the assumption that the electric charge is invariant under Lorentz transformations, so the charge density is altered by the length contraction of the capacitor plates but not in any other way. We can see how crucial this assumption is by comparing the transformations of electric and of gravitational fields. The source of electric fields is the electric charge, whereas the source of gravitational fields is the mass, or, more generally, the energy. In contrast to the electric charge, the energy is not invariant under Lorentz transformations. Because of this, the formulas (2) and (3) do not apply to gravitational fields, and neither do the other results given in this chapter. Both electric fields and gravitational fields are inverse-square force fields, but their relativistic properties are drastically different. We will discuss gravitational fields and Einstein's theory of gravitation in the next chapter.

If a moving point charge passes by a stationary point charge, the transformation equations (2) and (3) permit us to calculate the force experienced by the stationary charge. In a later section we will investigate the more general case in which both the charge that exerts the force and the charge that experiences the force are in motion.

As the following example illustrates, the enhancement of the perpendicular component of the electric field can be quite large if the velocity of the moving point charge is large.

■ **EXAMPLE 1.** A proton of energy 1 TeV passes by the nucleus of an iron atom at a distance of 10^{-10} m. Assume that the proton moves along a straight line with constant velocity. What is the magnitude of the electric field experienced by the nucleus at the instant of closest approach?

SOLUTION. In the rest frame of the proton, the Coulomb field of the proton at $r = 10^{-10}$ m has a magnitude

$$E = \frac{1}{4\pi\epsilon_0}\frac{e}{r^2} = 8.99 \times 10^9 \text{ F/m} \times \frac{1.60 \times 10^{-19}\text{ C}}{(10^{-10}\text{ m})^2} = 1.4 \times 10^{11} \text{ V/m} \qquad (4)$$

At the instant of closest approach, the electric field produced by the proton at the position of the iron nucleus is at right angles to the direction of motion, that is, the electric field is purely perpendicular. According to Eq. (3), when observed in the reference frame of the iron nucleus, this perpendicular field is enhanced by the factor $\gamma = 1/\sqrt{1 - V^2/c^2}$. Since the energy of the proton can be expressed as $E = mc^2\gamma$, the factor γ is simply $E/mc^2 \simeq 1000$ GeV/0.938 GeV $= 1.1 \times 10^3$. Multiplying the above electric field by this factor, we obtain

$$E' = 1.4 \times 10^{11} \times 1.1 \times 10^3 \text{ V/m} = 1.6 \times 10^{14} \text{ V/m}.$$

6.2 THE RELATIVISTIC TRANSFORMATION OF FORCE

In Newtonian mechanics, force is absolute--as we saw in Section 1.4, if the force acting on a particle in one inertial reference frame is **F**, then the force in any other inertial reference frame is exactly the same, that is, $\mathbf{F}' = \mathbf{F}$. In relativistic mechanics, the force is not absolute--the magnitude and the direction of the force changes as we change the reference frame. In this section, we obtain the relativistic transformation equations for the force, which we will need to investigate the behavior of electromagnetic forces when the charges have some general motion.

From Section 4.1 we know that the relativistic equation of motion in an inertial reference frame is $d\mathbf{p}/dt = \mathbf{F}$, and under a transformation to another inertial reference frame it retains the same form, $d\mathbf{p}'/dt = \mathbf{F}'$. Accordingly, the behavior of the force under a Lorentz transformation must be the same as the behavior of $d\mathbf{p}/dt$. To find the transformation equation for the force, we therefore first express $\mathbf{F}'$ as $d\mathbf{p}'/dt'$, then transform $d\mathbf{p}'$ and dt' into $d\mathbf{p}$ and dt, and finally express this in terms of **F**. The calculation is somewhat tedious and not very interesting, and an impatient reader may want to skip to the final results for the transformation of the components of the force, Eqs. (16) - (18).

The transformation of dt' is given by Eq. (2.17), from which:

$$dt = \gamma(dt' + V dx'/c^2) = \gamma dt'(1 + V v'_x/c^2)$$

or

$$dt' = \frac{dt}{\gamma(1 + Vv'_x/c^2)} \tag{5}$$

The transformations of the components of $d\mathbf{p}'$ are given by Eqs. (4.37) - (4.39), from which

$$\frac{dp'_x}{dt'} = \frac{\gamma(dp_x - VdE/c^2)}{dt'} = \gamma\left(\frac{dp_x}{dt} - \frac{V}{c^2}\frac{dE}{dt}\right) \times \gamma(1 + Vv'_x/c^2) \tag{6}$$

$$\frac{dp'_y}{dt'} = \frac{dp_y}{dt} \times \gamma(1 + Vv'_x/c^2) \tag{7}$$

$$\frac{dp'_z}{dt'} = \frac{dp_z}{dt} \times \gamma(1 + Vv'_x/c^2) \tag{8}$$

In Eq. (6), the derivative of the momentum equals the force, $dp_x/dt = F_x$, and the derivative of the energy equals the power, $dE/dt = \mathbf{F} \cdot \mathbf{v}$, so we obtain

$$F'_x = \gamma^2\left(F_x - \frac{V}{c^2}\mathbf{F} \cdot \mathbf{v}\right)(1 + Vv'_x/c^2) \tag{9}$$

For our purposes, it is desirable to express the velocity $\mathbf{v}$ in terms of $\mathbf{v}'$, since this will tell us how the force $\mathbf{F}'$ in the new reference frame depends on the velocity of the particle in that frame. The transformation equations for the velocity are Eq. (2.37) for v_x and corresponding equations for v_y and v_z:

$$v_x = \frac{v'_x + V}{1 + Vv'_x/c^2} \tag{10}$$

$$v_y = \frac{v'_y}{\gamma(1 + Vv'_x/c^2)} \tag{11}$$

$$v_z = \frac{v'_z}{\gamma(1 + Vv'_x/c^2)} \tag{12}$$

Consequently,

$$\mathbf{F} \cdot \mathbf{v} = F_x v_x + F_y v_y + F_z v_z$$

$$= \frac{1}{1 + V v'_x / c^2} [F_x (v'_x + V) + F_y v'_y / \gamma + F_z v'_z / \gamma] \quad (13)$$

By adding and subtracting $F_x v'_x / \gamma$ inside the brackets in this equation, we can rewrite it as

$$\mathbf{F} \cdot \mathbf{v} = \frac{1}{1 + V v'_x / c^2} [\mathbf{F} \cdot \mathbf{v}' / \gamma - F_x v'_x / \gamma + F_x \; (v'_x + V)] \quad (14)$$

Inserting this into Eq. (9), we obtain

$$F'_x = \gamma^2 \left(F_x - \frac{V}{\gamma c^2} \mathbf{F} \cdot \mathbf{v}' + F_x \frac{V v'_x}{\gamma c^2} - F_x \frac{V^2}{c^2} \right) \quad (15)$$

or, since $1 - V^2/c^2 = 1/\gamma^2$,

$$\boxed{F'_x = F_x - \frac{\gamma V}{c^2} \mathbf{F} \cdot \mathbf{v}' + \frac{\gamma V v'_x}{c^2} F_x} \quad (16)$$

Likewise, from Eqs (7) and (8) we immediately obtain

$$\boxed{F'_y = \gamma \left(1 + \frac{V v'_x}{c^2} \right) F_y} \quad (17)$$

and

$$\boxed{F'_z = \gamma \left(1 + \frac{V v'_x}{c^2} \right) F_z} \quad (18)$$

These expressions show that the transformation of the x component of the force differs from that of the y and z components. Hence both the magnitude and the direction of the force change when we transform from one reference frame to another. Furthermore, the force $\mathbf{F}'$ depends not only on the relative velocity $\mathbf{V}$ of the reference frames but also on the velocity $\mathbf{v}'$ of the particle in the new reference frame, that is, the force is velocity-dependent.

Of course, for low velocities V and v', the relativistic formulas (16) - (18) for the components of $\mathbf{F}'$ agree with the Newtonian formulas. Low velocities mean $V \ll c$ and $v' \ll c$, so Vv'_x/c^2 is negligible and $\gamma \simeq 1$. The relativistic formulas (16) - (18) then reduce to $F'_x \simeq F_x$, $F'_y \simeq F_y$, and $F'_z \simeq F_z$, which are indeed the Newtonian formulas. Note that the deviations between the relativistic and the Newtonian formulas are quadratic in the velocity, that is, they are proportional to the quantities V^2/c^2 and Vv'/c^2, with two factors of the velocity in the numerator and two factors of c in the denominator.

6.3 THE MAGNETIC FORCE AND FIELD*

From our formula for the transformation of the force, we can confirm our previous result for the transformation of the parallel and perpendicular components of the electric field. For this purpose, suppose that in the first reference frame there is an electric field **E**, so the force on a charged particle is $\mathbf{F} = q\mathbf{E}$. Suppose that this charged particle is at rest in the second reference frame, so $\mathbf{v}' = 0$. Then the terms involving v'_x disappear from Eqs. (16) - (18) and these equations reduce to

$$F'_x = F_x \qquad (19)$$

$$F'_y = \gamma F_y \qquad (20)$$

$$F'_z = \gamma F_z \qquad (21)$$

Since F_x is the component of the force parallel to the velocity **V**, and since F_y and F_z are the components perpendicular, we can write these equations as

$$\mathbf{F}'_{\parallel} = \mathbf{F}_{\parallel} \qquad (22)$$

$$\mathbf{F}'_{\perp} = \gamma \mathbf{F}_{\perp} \qquad (23)$$

With $\mathbf{F} = q\mathbf{E}$ and $\mathbf{F}' = q\mathbf{E}'$, these equations immediately give us

$$\mathbf{E}'_{\parallel} = \mathbf{E}_{\parallel} \qquad (24)$$

$$\mathbf{E}'_{\perp} = \gamma \mathbf{E}_{\perp} \qquad (25)$$

which confirms our previous results (2) and (3). Note that this argument, again, hinges on the invariance of the electric charge under Lorentz transformations, that is, the value of the electric charge q is

---

* The derivations of the fields of uniformly moving charges and accelerated charges presented in this and the next section are simplified versions of derivations presented by Frisch and Wilets [1]. A more rigorous treatment is given in their paper. An additional assumption stated and discussed there is that a rate of change of the acceleration produces no extra terms in the radiation fields.

the same in the old and the new reference frames.

Next, we want to examine what happens if the charged particle is not at rest in the new reference frame, so $\mathbf{v}' \neq 0$. Then we have to consider the terms involving v'_x in Eqs. (16) - (18). In F'_x, F'_y, and F'_z these extra terms are, respectively,

$$-\frac{\gamma V}{c^2}\mathbf{F}\cdot\mathbf{v}' + \frac{\gamma V v'_x}{c^2}F_x \, , \quad \frac{\gamma V v'_x}{c^2}F_y \, , \quad \frac{\gamma V v'_x}{c^2}F_z$$

But these three quantities are the x, y, and z components of the vector

$$-\frac{\gamma \mathbf{V}}{c^2}\mathbf{F}\cdot\mathbf{v}' + \frac{\gamma \mathbf{V}\cdot\mathbf{v}'}{c^2}\mathbf{F} \tag{26}$$

This can be easily verified by checking the x, y, and z components of the expression (26), taking into account that $\mathbf{V}$ has only an x component, with $V_x = V$, so $\mathbf{V}\cdot\mathbf{v}' = V_x v'_x = V v'_x$. With $\mathbf{F} = q\mathbf{E}$, the extra, velocity-dependent force (26) is therefore

$$-\frac{q\gamma \mathbf{V}}{c^2}\mathbf{E}\cdot\mathbf{v}' + \frac{q\gamma \mathbf{V}\cdot\mathbf{v}'}{c^2}\mathbf{E} \tag{27}$$

As a final step in our calculation, we exploit the dot product formula for the cross product of three vectors:

$$\mathbf{A}\times(\mathbf{B}\times\mathbf{C}) = (\mathbf{A}\cdot\mathbf{C})\,\mathbf{B} - (\mathbf{A}\cdot\mathbf{B})\,\mathbf{C} \tag{28}$$

The two terms in the expression (27) are of the same form as the two terms on the right side of this dot product formula, and accordingly we can rewrite (27) as a cross product of three vectors:

$$\frac{q\gamma}{c^2}\mathbf{v}'\times(\mathbf{E}\times\mathbf{V}) \tag{29}$$

This force has a direction perpendicular to the velocity $\mathbf{v}'$. Thus, it has the character of a magnetic force exerted by a magnetic field. From the study of magnetism, we know that the force exerted by a magnetic field $\mathbf{B}'$ on a particle of charge q moving with velocity $\mathbf{v}'$ is

$$q\,\mathbf{v}'\times\mathbf{B}' \tag{30}$$

Comparing the expressions (29) and (30), we see that the magnetic field in the new reference frame is

$$\mathbf{B}' = \frac{\gamma}{c^2} \mathbf{E} \times \mathbf{V} \tag{31}$$

Equations (24), (25) and (31) are the equations for the transformation of an electric field in one reference frame into electric and magnetic fields in another reference frame. These equations state that an electric field in one inertial reference frame transforms into a combination of electric and magnetic fields in another inertial reference frame.

Equation (31) gives us a quick way to calculate the magnetic field of a current distribution, provided that this current distribution can be regarded as a purely translational motion of a given distribution of electric charge. Examples of such current distributions include a uniform current flowing along an infinitely long straight wire and a uniform current flowing along an infinitely long straight ribbon. These current distributions can be regarded as a translational motion of a charged line and of a charged ribbon along its long direction.

■ **EXAMPLE 2.** An infinitely long straight horizontal line of charge has a uniform charge distribution of λ coulombs per unit length. The electric field of this line of charge can be easily calculated by integration, starting with Coulomb's Law for the individual charge elements on the line. At a distance r from the line, the electric field has a radial direction, perpendicular to the line of charge (see Fig. 6.6a), and a magnitude

$$E = \frac{1}{2\pi\epsilon_0} \frac{\lambda}{r} \tag{32}$$

Consider this line of charge in a new reference frame moving parallel to the line of charge with velocity $\mathbf{V}$ toward the left. In this new reference frame, the line of charge has a velocity $-\mathbf{V}$ toward the right, and this moving line of charge constitutes a current. (a) Calculate the magnetic field in the new reference frame by means of the transformation equation (31). (b) Express this magnetic field in terms of the current in the new reference frame, and verify that the result is consistent with Ampère's Law.

SOLUTION. (a) According to Eq. (31), the magnetic field $\mathbf{B}'$ is perpendicular to the electric field $\mathbf{E}$ and to the velocity, that is, the magnetic field is tangent to a concentric circle around the line of charge, and the magnetic field lines are concentric circles around the direction of motion (see Fig. 6.6b). The magnitude of the magnetic field is

$$B' = \frac{\gamma}{c^2} \frac{1}{2\pi\epsilon_0} \frac{\lambda V}{r} \tag{33}$$

To express this in a more familiar form, we note that in the new reference frame, the line of charge is length-contracted, and its charge per unit length is larger than

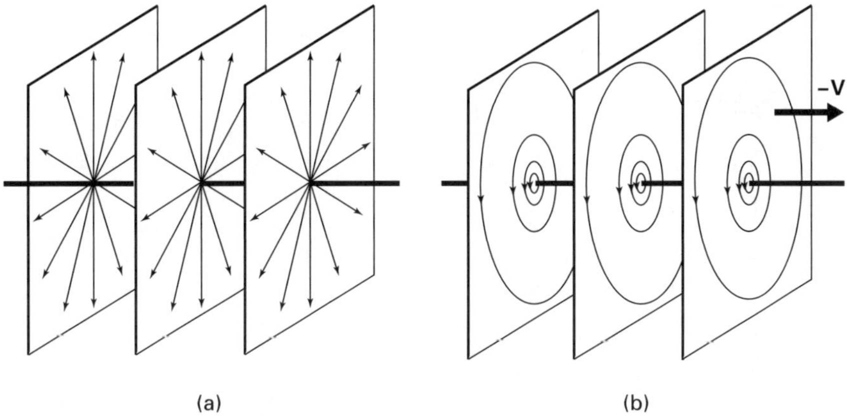

Fig. 6.6 (a) Electric field **E** of an infinitely long straight line of charge at rest. The electric field has a radial direction. (b) The line of charge moves toward the right relative to the new reference frame. The magnetic field **B′** is perpendicular to the electric field and to the velocity. The magnetic field lines are concentric circles around the line of charge.

the charge per unit length in the old reference frame by a factor of γ,

$$\lambda' = \gamma\lambda \tag{34}$$

Furthermore, we recall that $c^2 = 1/\epsilon_0\mu_0$. Substituting this into Eq. (33), we find that

$$B' = \frac{\mu_0}{2\pi}\frac{\lambda' V}{r} \tag{35}$$

(b) The product of the charge density λ' and the velocity V of the line of charge is the current,

$$I' = \lambda' V \tag{36}$$

Hence the magnetic field is simply

$$B' = \frac{\mu_0}{2\pi}\frac{I'}{r} \tag{37}$$

This agrees with the familiar expression given by Ampère's Law for the magnetic field of a long, straight wire carrying a current I'. Thus, the relativistic transformation formula (31) yields a basic result of magnetism. The calculation of the magnetic field by this method is trivial but requires that we know the electric field (32).

Note that the expression (34) for the charge per unit length hinges on the relativistic length contraction. Without the length contraction, we would obtain a result different from Eq. (37), that is, we would obtain a result inconsistent with Ampère's

Law. We can therefore regard the agreement between Eq. (37) and Ampère's Law as indirect evidence for the length contraction.

From the formula (31) for the transformation of the electric field we can also immediately deduce the magnetic field of a uniformly moving point charge. In its own rest frame, the charge has a radial electric field of magnitude $E = q/4\pi\epsilon_0 r^2$. In a reference frame moving toward the left with velocity $\mathbf{V}$, the charge moves toward the right with velocity $\mathbf{v}' = -\mathbf{V}$, and according to Eq. (31), the magnetic field observed in the new reference frame is

$$\mathbf{B}' = \frac{\gamma}{c^2}\mathbf{E} \times \mathbf{V} \tag{38}$$

or, in terms of the velocity of the charge,

$$\mathbf{B}' = \frac{\gamma}{c^2}\mathbf{v}' \times \mathbf{E} \tag{39}$$

Since the cross product of two vectors depends only on their mutually perpendicular components, we can replace $\mathbf{E}$ on the right side of this equation by $\mathbf{E}_\perp$. Furthermore, the product $\gamma\mathbf{E}_\perp$ equals $\mathbf{E}'_\perp$. This permits us to express Eq. (39) as

$$\mathbf{B}' = \frac{1}{c^2}\mathbf{v}' \times \mathbf{E}'_\perp$$

Here we can replace $\mathbf{E}'_\perp$ on the right side by $\mathbf{E}'$, so

$$\mathbf{B}' = \frac{1}{c^2}\mathbf{v}' \times \mathbf{E}' \tag{40}$$

This is a simple expression for the magnetic field of a moving charge in terms of its electric field, with both fields reckoned in the reference frame in which the charge has a velocity $\mathbf{v}'$. Evidently, the electric field $\mathbf{E}'$ appearing in Eq. (40) is the Lorentz-contracted electric field of the moving charge, with field lines as indicated in Fig. 6.2.

Since the electric field has a radial direction away from the charge, the cross product tells us that the magnetic field is tangent to a circle concentric with the direction of motion, that is, the magnetic field lines are concentric circles around the direction of motion (see Fig. 6.7).

■ **EXAMPLE 3.** In Example 1 we calculated the electric field produced by a moving proton at the instant of its closest approach to an iron nucleus. What is the magnetic field produced by this proton at this instant?

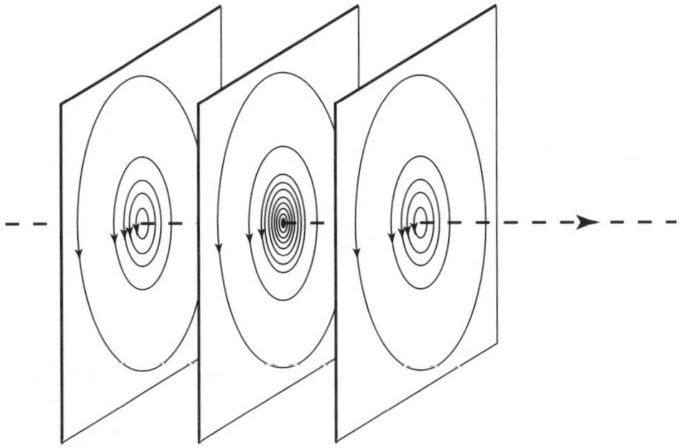

Fig. 6.7 Magnetic field lines for the magnetic field of a positive point charge moving toward the right. The field lines are concentric circles around the line of motion. The density of field lines becomes infinite near the point charge (at the center of the diagram), and the density is a decreasing function in all directions away from the point charge. On the line of motion, ahead and behind the point charge, the density of field lines is zero.

SOLUTION. Since the energy of the proton is very large, the velocity of the proton is $v' \simeq c$. At the position of the nucleus, at the instant of closest approach, the electric field $\mathbf{E}'$ is perpendicular to the velocity. Hence, according to Eq. (40), the magnetic field is

$$B' = \frac{1}{c^2} v'E' \simeq \frac{1}{c} E' \tag{41}$$

$$= \frac{1}{c} \times 1.6 \times 10^{14} \text{ V/m} = 5.3 \times 10^5 \text{ T}$$

The direction of this magnetic field is at right angles to the directions of the velocity and the electric field.

Note that such perpendicular electric and magnetic fields, with magnitudes related by $B' = E'/c$, are characteristic of electromagnetic waves. Thus, the electric and magnetic fields of the proton passing by the nucleus at a speed near c mimic the fields of an electromagnetic wave. The effect of this pulse of electric and magnetic fields on the iron nucleus is nearly the same as the effect of a genuine electromagnetic wave pulse.

From a knowledge of the magnetic field of a moving charge, we can, in principle, calculate the magnetic field of any arbitrary distribution of currents, since any such distribution consists of some collection of moving charges. Hence, the Biot-Savart Law and Ampère's Law are implicit in Eq. (40). In fact, Eq. (40) goes somewhat beyond Ampère's Law, describing not only static magnetic fields but also magnetic fields with some time dependence. However, all the electric and magnetic fields we have investigated so far are those of charges with uniform motion. Thus, charges with accelerated motion are excluded, and this imposes some restrictions on the kinds of electric and magnetic fields that we can construct. In the next section we will deal with the fields of accelerated charges. This will remove the restrictions, and it permits us to construct all the kinds of electric and magnetic fields consistent with Maxwell's equations.

6.4 THE GENERAL TRANSFORMATION LAWS FOR E AND B

According to Eqs. (24), (25), and (31), an electric field in one reference frame becomes a combination of electric and magnetic fields in another reference frame. But the converse is also true: a magnetic field in one reference frame becomes a combination of magnetic and electric fields in another reference frame. To see how this comes about, suppose that in the old reference frame there is a magnetic field **B** which exerts a magnetic force $\mathbf{F} = q\,\mathbf{v} \times \mathbf{B}$ on a charged particle. To discover the electric field in the new reference frame, we assume that the particle is at rest in this new reference frame (the electric force is the force that a particle experiences when at rest, with $\mathbf{v}' = 0$). If so, then the velocity in the old reference frame will be $\mathbf{v} = \mathbf{V}$, and $\mathbf{F} = q\mathbf{V} \times \mathbf{B}$. The transformed force $\mathbf{F}'$ can be most conveniently calculated from Eqs. (22) and (23), which are valid for $\mathbf{v}' = 0$. Since the magnetic force $\mathbf{F} = q\mathbf{V} \times \mathbf{B}$ is perpendicular to the velocity, the parallel component in Eq. (22) is zero, and only the perpendicular component in Eq. (23) remains:

$$\mathbf{F}' = q\,\gamma\mathbf{V} \times \mathbf{B} \tag{42}$$

This tells us that the electric field in the new reference frame is

$$\mathbf{E}' = \gamma\mathbf{V} \times \mathbf{B} \tag{43}$$

This has the character of a motional electric field, familiar from the study of electromagnetism. The direction of this electric field is perpendicular to the velocity **V**.

If there is not only a magnetic field **B** in the old reference frame but also an electric field **E**, then the net electric field in the new reference frame must be calculated by combining the transformations (24) and (25) for the electric-field components and the transformation for (43) for the magnetic field. The net result is then

$$\boxed{\mathbf{E}'_{\parallel} = \mathbf{E}_{\parallel}} \tag{44}$$

$$\boxed{\mathbf{E}'_{\perp} = \gamma \mathbf{E}_{\perp} + \gamma \mathbf{V} \times \mathbf{B}} \tag{45}$$

These are the general transformation equations for the electric field in the presence of a magnetic field.

There is also a corresponding equation for the transformation of the magnetic field in the presence of an electric field. We have found one part of that transformation in Eq. (31), but we did not investigate how the parallel and the perpendicular components of the magnetic field transform. The complete formula for the transformation of the net magnetic field can be obtained by more manipulation of the force transformation law. The final result is

$$\boxed{\mathbf{B}'_{\parallel} = \mathbf{B}_{\parallel}} \tag{46}$$

$$\boxed{\mathbf{B}'_{\perp} = \gamma \mathbf{B}_{\perp} + \frac{\gamma}{c^2} \mathbf{E} \times \mathbf{V}} \tag{47}$$

Note the strong similarity between Eqs. (44), (45) and Eqs. (46), (47).

The transformation formulas for electric and magnetic fields reveal a close kinship between these fields. Paraphrasing Minkowski's statement about space and time, we can say, "henceforth the electric field by itself, and the magnetic field by itself, are doomed to fade away into mere shadows, and only a union of the two will preserve an independent reality." Einstein's discovery of the relativistic relationships between electric and magnetic fields established that electromagnetism is a unified field theory. Electric and magnetic fields are merely two aspects of a single entity, called the electromagnetic field tensor, just as space and time are two aspects of a single spacetime. But the study of the four-dimensional tensor formulation of Maxwell's equations requires some knowledge of tensor analysis, and we cannot deal with this here.

*6.5 THE ELECTRIC FIELD OF AN ACCELERATED CHARGE**

In Sections 6.1 and 6.4 we exploited relativity to obtain the electric and magnetic fields of a point charge in uniform motion. Now we want to obtain the fields of an accelerated point charge. Besides the velocity-dependent fields we found above, such an accelerated charge has extra acceleration-dependent fields, or **radiation fields**. These

* The derivation of the electric field of an accelerated charge given in this section was originally formulated by J. J. Thomson [2] around 1900. It has been rediscovered several times since then.

radiation fields are electric and magnetic disturbances that spread outward from the accelerated charge, forming a spherical wavefront, like the shock wave from the explosion of a fire cracker. If the acceleration of the charge is periodic (back-and-forth motion), then the radiation fields form a periodic electromagnetic wave. Light is a special case of such an electromagnetic disturbance, and we know that the speed of propagation of this special kind of electromagnetic disturbance is c, the speed of light. We will therefore make the natural assumption that the speed of propagation of all disturbances in electric and magnetic fields is the speed of light.

To derive a formula for the radiation field produced by the acceleration of a point charge, we assume that the charge is initially at rest, then accelerates for a short time δt with acceleration a, and then continues with uniform motion at its final velocity $a \times \delta t$. Assuming that the acceleration begins at time $t = 0$, we can summarize the motion by the scheme

$t < 0$	charge is at rest
$0 < t < \delta t$	charge has acceleration a
$t > \delta t$	charge moves with uniform velocity $v = a\delta t$

Figure 6.8 shows the initial electric field lines of the charge, when the charge is at rest at O. Figure 6.9 shows the field lines at a time t much later than δt, when the charge is instantaneously at P. The field lines in the immediate vicinity of the charge are the usual field lines for a charge in uniform motion; they diverge radially from the present position P of the charge. But the field lines far from the charge are the old field lines that the charge had when it still was at rest at its initial position O. These field lines still diverge radially from the initial position O. The field lines far from the charge have not yet had time to change--the electric disturbance emitted during the interval of acceleration has not yet reached the regions far from the charge. The electric disturbance takes the form of a kink in the field lines that connects the old and the new lines. This kink travels outward at the speed of light and gradually deforms the old pattern of field lines into the new pattern.

Since the kink, like any other electric disturbance, propagates at speed c, it will reach a radial distance $r = ct$ at time t. Thus, the sphere of this radius is the boundary between the old and the new field lines. The kink contains the radiation field we seek. In the region of the kink, the electric field has a longitudinal component (along the direction of the radius, or the direction of propagation) and also a transverse component (at right angles to the direction of propagation). The longitudinal component is simply the Coulomb field, whereas the transverse component is the radiation field. The transverse direction distinguishes the radiation field from the Coulomb field--it is the characteristic feature of the radiation field.

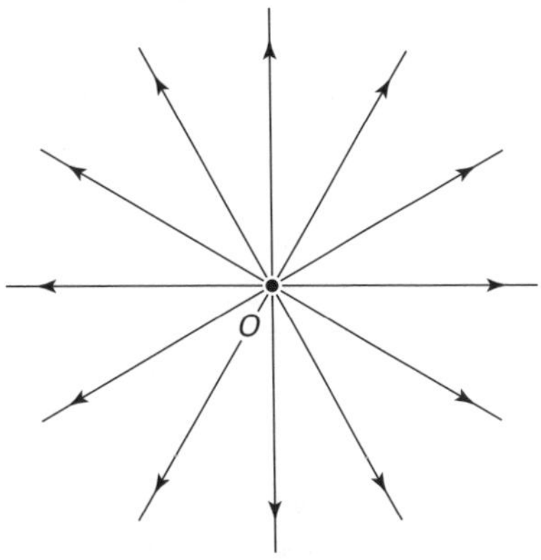

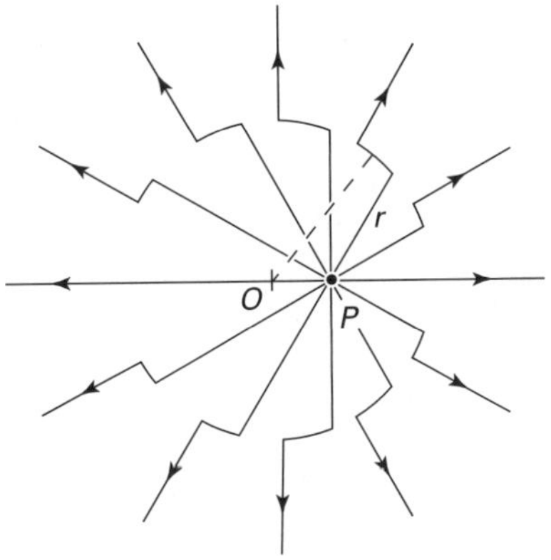

Fig. 6.8 (Left) Electric field lines for a charge initially at rest at the point O.

Fig. 6.9 (Right) Electric field lines for the charge at a time t after it has been subjected to a brief acceleration lasting for a time δt. The initial position of the charge is O, and the present position of the charge is P. In this diagram, it has been assumed that $t >> \delta t$. On the scale of this diagram, the distance over which the particle accelerates is then not noticeable, and the diagram is drawn as though the acceleration occurred instantly at P. Between O and P the charge has therefore moved with uniform velocity. The distance OP is $vt = (a\delta t)t$, and the radial distance to the kink in the field lines is $r = ct$.

Figure 6.10 shows the electric field lines in one sector in more detail. By inspection of this figure, we can calculate the strength of the transverse radiation field. The field lines in this sector span an angle $\delta\theta$ at the charge, and this angular width has been selected to match the angular width of the kink in a field line. All these field lines in this sector bend toward the left, and they cross a radial line OS within the small interval $\delta r = c\delta t$ marked in the figure. The density of field lines at this location tells us the strength of the transverse electric field. To calculate this strength, it is convenient to begin with the known strength of the electric Coulomb field at the boundary of the sector, at radius r,

$$E = \frac{1}{4\pi\epsilon_0}\frac{q}{r^2} \tag{48}$$

and then multiply this strength by a factor that represents the increase of density of the lines in the kink. Just before they enter the kink, the radial field lines of the sector are spread over an arc $r\delta\theta$; in the kink, the transvere field lines are spread over a width $\delta r = c\delta t$. Hence the transverse electric field is stronger than the Coulomb field by a factor $r\delta\theta / c\delta t$,

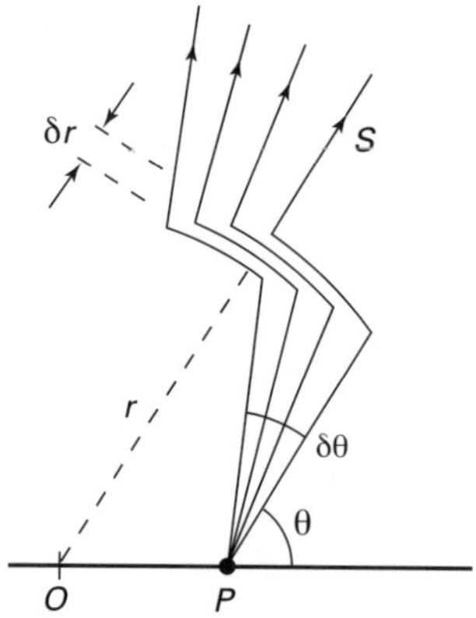

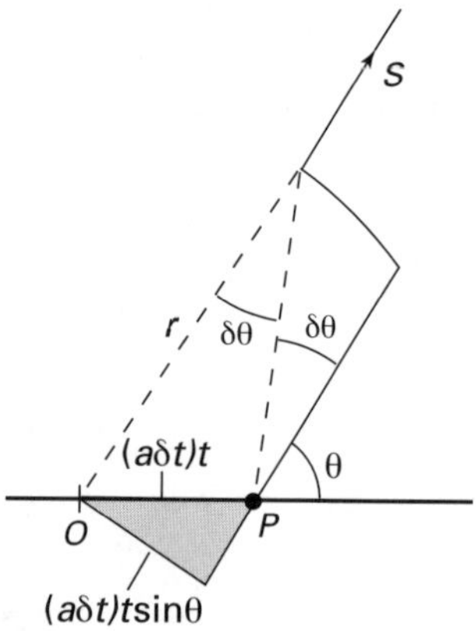

Fig. 6.10 (Left) Detailed view of field lines in a sector of angular width $\delta\theta$ equal to the angular width of the kink in the field line PS.

Fig. 6.11 (Right) Geometrical relationships between the kinked field line PS and the distance traveled by the particle, $OP = (a\delta t)t$. For the sake of clarity, the distance traveled by the particle and the shaded triangle have been shown larger than they are under nonrelativistic conditions.

$$E_{\text{rad}} = \frac{1}{4\pi\epsilon_0}\frac{q}{r^2} \times \frac{r\delta\theta}{c\delta t} \tag{49}$$

To evaluate $\delta\theta$, we use Fig. 6.11, which shows the geometry. The small shaded triangle has as hypotenuse the distance traveled by the point charge, that is, $(a\delta t)t = (a\delta t)r/c$; hence its opposite side is $(a\delta t)r/c \times \sin\theta$. Under the assumption that the speed of the particle is nonrelativistic ($v << c$), the sides of the shaded triangle are much smaller than r, and the angle $\delta\theta$ is small. The angle $\delta\theta$ is then approximately equal to the opposite side divided by r, that is, $\delta\theta \simeq (a\delta t)\sin\theta/c$. Substituting this into Eq. (49), we find

$$E_{\text{rad}} = \frac{1}{4\pi\epsilon_0}\frac{q}{r^2} \times \frac{r(a\delta t)\sin\theta}{c^2\delta t} \tag{50}$$

or

$$\boxed{E_{\text{rad}} = \frac{1}{4\pi\epsilon_0}\frac{qa\sin\theta}{c^2 r}} \tag{51}$$

This is the transverse radiation field of the point charge. Note that this field is directly proportional to the acceleration. Also note that it decreases in inverse proportion to the distance, in contrast to the Coulomb field, which decreases in inverse proportion to square of the distance. Thus, the radiation field decreases less sharply with distance than the Coulomb field, and it remains significant at large distances, where the Coulomb field practically disappears.

The radiation field produced by an acceleration a acting at some given time reaches the distance r after a time delay r/c; the field needs this time delay to propagate over the distance r. This means that the effect of the acceleration will be felt at a distance r at a delayed time. We can incorporate this delay into our equation by expressing the acceleration as a function $a(t)$ of time and expressing the radiation field as a function of the corresponding delayed time:

$$E_{\text{rad}}(t + r/c) = \frac{1}{4\pi\epsilon_0}\frac{q\sin\theta}{c^2 r}\,a(t) \tag{52}$$

or, equivalently,

$$E_{\text{rad}}(t) = \frac{1}{4\pi\epsilon_0}\frac{q\sin\theta}{c^2 r}\,a(t - r/c) \tag{53}$$

This equation indicates how the value of the electric field at time t is related to the acceleration at time $t - r/c$, the retarded time. From this equation, we can calculate the radiation field generated by a charge with some arbitrary, variable acceleration. For instance, we can calculate the radiation field emitted by a charge moving back and forth with simple harmonic motion. Such a charge generates a harmonic wave with regularly spaced concentric wavefronts.

Besides the electric radiation field, there is also a magnetic radiation field. The existence of this magnetic radiation field can be deduced by examining the behavior of the electric radiation field under Lorentz transformations. Suppose we repeat the calculation of the electric radiation field in a reference frame that has a velocity **V** relative to the reference frame used in the calculation that led to Eq. (53). In this reference frame, we can again calculate the electric radiation field by inspecting the kink in the electric field lines. As in our previous calculation, the electric radiation field is transverse to the direction of propagation of the kink. But this transversality of the radiation field in the new and the old reference frames leads to a paradox when we compare the directions of propagation of some portion of the radiation field (or some portion of the wavefront) in the new and the old reference frames. For instance, consider the portion of the wavefront that, in the old reference frame, propagates upward, at right angles to the direction of the acceleration. The electric radiation field in this part of the wavefront is horizontal (see Fig. 6.12a). Next, consider the same portion of the wavefront as observed in the new reference frame with a velocity **V** toward the left. In this new refer-

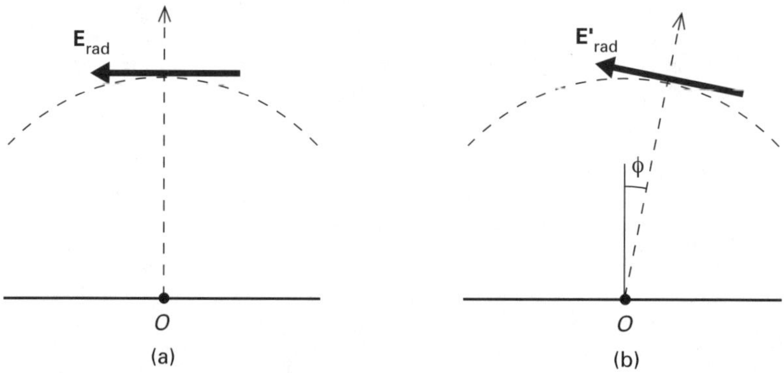

Fig. 6.12 (a) Direction of the electric radiation field $\mathbf{E}_{rad}$ in the portion of the wavefront that propagates upward, at right angles to the direction of the acceleration ($\theta = 90^0$). (b) As seen in a reference frame moving toward the left, the direction of propagation of the wavefront is tilted toward the right, and the direction of the electric radiation field $\mathbf{E}'_{rad}$ is also tilted toward the right.

ence frame, the direction of propagation of the wave is tilted toward the right, by the usual aberration effect. The angle of this tilt is given by the aberration formula $\phi = \tan^{-1}(\gamma V/c)$ (see Problem 2.30). Since the electric radiation field is always perpendicular to the direction of propagation, the direction of the electric radiation field will be tilted toward the right by the same angle (see Fig. 6.12b). But this tilt of the electric radiation field conflicts with the transformation laws (2) and (3) for electric fields. *In the absence of a magnetic field*, the transformation laws (2) and (3) assert that if the electric field in the first reference frame is entirely parallel to the velocity then the electric field in the second reference frame is necessarily also entirely parallel. Thus, the tilted electric field in Fig. 6.12b contradicts the transformation law!

To resolve this paradox, we have to invoke the existence of a magnetic field. We have to assume that the electric radiation field E_{rad} in the first reference frame is accompanied by a magnetic radiation field B_{rad}, so that the extra magnetic term in the transformation law (45) provides us with an extra electric field $\gamma \mathbf{V} \times \mathbf{B}_{rad}$ in the second reference frame. Evidently, this extra electric field must be directed in the vertical upward direction, so that the sum of the parallel electric field and the extra vertical electric field gives the tilted electric field shown in Fig. 6.12b. It is easy to see that the magnetic radiation field required to accomplish this must be at right angles to the electric radiation field and to the direction of propagation and that it must have a magnitude

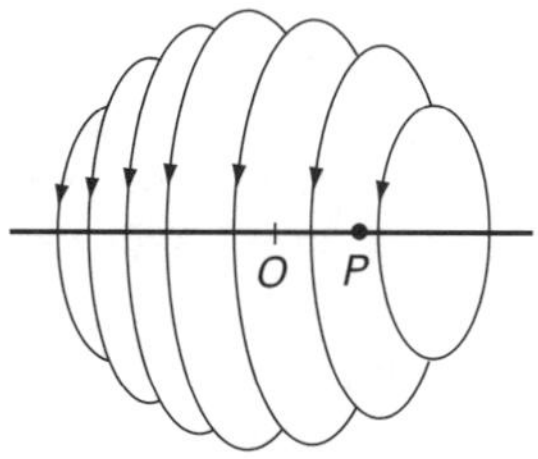

Fig. 6.13 Field lines of the magnetic radiation field $\mathbf{B}_{rad}$. The field lines are circles concentric with the direction of the acceleration of the point charge. This magnetic radiation field exists only at the spherical wavefront, where there is an electric radiation field. In contrast to Fig. 6.7, the field lines do not fill all of space (unless the acceleration lasts indefinitely).

$$\boxed{B_{rad} = \frac{E_{rad}}{c}} \tag{54}$$

The field lines of the magnetic radiation field are concentric circles around the direction of the acceleration (see Fig. 6.13).

With these results for the electric and magnetic radiation fields, we have accomplished our program of constructing electromagnetism from relativity. The formulas for the fields of a uniformly moving charge and the fields of an accelerated charge permit us to calculate the fields generated by any collection of point charges with any kind of motion. This means we can construct every kind of electric and magnetic field that any kind of charge distribution can produce, and this means that we can generate every conceivable solution of the Maxwell equations. But knowledge of all solutions of a system of equations is effectively equivalent to knowledge of the equations themselves. Thus, by combining Coulomb's Law with relativity, we can deduce all of the Maxwell equations.

REFERENCES

1. D. H. Frisch and L. Wilets, Am. J. Phys. **24**, 574 (1955).

2. J. J. Thomson, *The Corpuscular Theory of Matter* (Charles Scribner's Sons, New York, 1907), p. 45.

PROBLEMS

1. In Example 1 we calculated the electric field produced by a moving proton in the reference frame of an iron nucleus at the instant of closest approach. At the same instant, what is the electric field generated by the iron nucleus in the refer-

ence frame of the proton?

2. Consider a point charge q that is moving at a uniform velocity v relative to you. Find a formula for the magnitude of the electric field at a given fixed distance r from the point charge, measured in your reference frame. Express the answer as a function of the angle θ between the radial line and the direction of motion.

3. Show that if the force is parallel to the velocity **V** of the Lorentz transformation, so $F_y = 0$ and $F_z = 0$, then the force transformation law yields $F'_x = F_x$, that is, a parallel force is invariant.

4. (a) Show that the three equations (16) - (18) can be written as a single vector equation:

$$\mathbf{F}' = \frac{\mathbf{F}\cdot\mathbf{V}}{V^2}\mathbf{V} + \gamma F - \frac{\gamma(\mathbf{F}\cdot\mathbf{V})}{V^2}\mathbf{V} + \frac{\gamma(\mathbf{V}\cdot\mathbf{v}')}{c^2}\mathbf{F} - \frac{\gamma(\mathbf{F}\cdot\mathbf{v}')}{c^2}\mathbf{V}$$

(*Hint*: Note that **V** is a vector in the x direction, so its y and z components are zero, and therefore $\mathbf{F}\cdot\mathbf{V} = F_x V$ and $\mathbf{V}\cdot v' = Vv'_x$.)
(b) Show that the equation of part (a) can be rewritten as

$$\mathbf{F}' = \frac{\mathbf{F}\cdot\mathbf{V}}{V^2}\mathbf{V} + \frac{\gamma}{V^2}\mathbf{V}\times(\mathbf{F}\times\mathbf{V}) + \frac{\gamma}{c^2}\mathbf{v}'\times(\mathbf{F}\times\mathbf{V})$$

5. Equations (16) - (18) give the transformation laws for the components of the force in terms of the velocity **v′** in the new reference frame. Show that in terms of the velocity **v** in the the old reference frame, the transformation law takes the alternative form

$$F'_x = \frac{F_x - V/c^2\,\mathbf{F}\cdot\mathbf{v}}{1 - Vv_x/c^2}$$

$$F'_y = \frac{F_y}{\gamma(1 - Vv_x/c^2)}$$

$$F'_z = \frac{F_z}{\gamma(1 - Vv_x/c^2)}$$

6. In a hydrogen atom, the electron orbiting the nucleus moves through the electric field of the nucleus; hence, in the rest frame of the electron, there will be not only an electric field but also a magnetic field. Assume that the electron moves in a circular orbit of radius 0.529×10^{-10} m at a speed of $v \simeq c/137$. Calculate the electric field produced by the nucleus at the position of the electron, and calculate the magnetic field observed in the instantaneous rest frame of the electron.

7. A parallel-plate capacitor has plates with uniform charge distributions of $\pm\sigma$ coulombs per unit area. In the rest frame of the capacitor, the electric field between the plates is $E = \sigma/\epsilon_0$.
(a) In a reference frame moving with velocity V in a direction parallel to the plates, what is the electric field E'?
(b) What is the magnetic field B'?
(c) Show that the magnetic field B' agrees with the magnetic field generated by a pair of large plates carrying opposite uniform surface currents of $\gamma\sigma V$ amps per unit length measured perpendicular to **V**. (*Hint*: Use Ampère's Law to calculate the magnetic field between the plates from the current.)

8. A uniform charge distribution of σ coulombs/m^2 is placed on an infinitely long straight ribbon of width d. (a) Calculate the electric field at the point P in the plane of the ribbon, at a distance r from the midline. (b) From this electric field, deduce the magnetic field of a uniform current I flowing along such a straight ribbon.

9. A uniform charge distribution of σ coulombs/m^2 is placed on an infinitely long, straight ribbon of width d. (a) Calculate the electric field at the point P above the midline of the ribbon, at a distance r. (b) From this electric field, deduce the magnetic field of a uniform current I flowing along such a straight ribbon.

10. Suppose that in one reference frame, we have a uniform electric field in some direction and no magnetic field. Is it possible that because of the relativistic transformation law, the electric field in a second reference frame is zero? Is it possible that the magnetic field in a second reference frame is zero?

11. Show that $E^2 - c^2B^2$ is invariant under Lorentz transformations. This implies that if the magnitudes of **E** and c**B** are equal in one inertial reference frame, then they are equal in all other reference frames. The electric and magnetic fields of a light wave are an example of fields that satisfy this equality.

12. Show that **E** · **B** is invariant under Lorentz transformations. This implies that if the electric and magnetic fields are perpendicular in one reference frame, then they are perpendicular in all other reference frames. The electric and magnetic fields of a light wave are an example of fields that satisfy this condition of perpendicularity.

13. What are the field transformation laws inverse to (44) - (47)? That is, what are the expressions for **E** and **B** in terms of **E**$'$ and **B**$'$?

14. In practice, magnetic fields are usually produced by means of currents on electrically neutral wires, so in the rest frame of the wire, there is a magnetic field but no electric field. Consider such a pure magnetic field $B = \mu_0 I/2\pi r$, produced by a long straight wire carrying a current I. Suppose that a point charge q of mass m instantaneously moves parallel to this wire with velocity v at a distance r.
(a) What is the radial magnetic force that acts on the point charge and what is the

acceleration?

(b) In the rest frame of the point charge there is no magnetic force, but there is a radial electric field that produces a radial force. What is the radial electric field in this reference frame, what is the force on the point charge, and what is the acceleration?

(c) Since there is a radial electric field in this reference frame, there must be a charge per unit length on the wire in this reference frame. Deduce the charge per unit length from the electric field.

(d) Show that this charge per unit length can be explained by examining the positive and negative charges (ions and conduction electrons) that exist within the wire. In the rest frame of the wire, the ions are at rest, but the conduction electrons are in motion. The charge of the ions is λ per unit length, and the charge of the conduction electrons is $-\lambda$ per unit length, giving a net charge of zero; but the motion of the conduction electrons produces a current $I = \lambda v$ (for the sake of simplicity, assume that the speed of the conduction electrons is the same as the speed of the point charge). Show that as a result of the length contraction of the ionic and electronic charge distributions, the wire will have a net electric charge per unit length in the rest frame of the point charge.

15. The equation of motion of a charged particle in uniform perpendicular electric and magnetic fields, or **crossed fields**, is complicated and difficult to solve. However, the solution can be simplified by performing a Lorentz transformation to a new reference frame moving perpendicularly to the plane of **E** and **B**. By an appropriate choice of the magnitude of the Lorentz velocity, the electric field or the magnetic field can be reduced to zero by such a transformation. If $E < cB$, then the electric field can be reduced to zero, and if $E > cB$, then the magnetic field can be reduced to zero. In this problem we will deal with the first case (the next problem deals with the second case).

Under the assumption $E < cB$, take the velocity **V** of the Lorentz transformation as

$$\mathbf{V} = \frac{\mathbf{E} \times \mathbf{B}}{B^2}$$

(a) Show that $V < c$.

(b) Show that with this velocity **V**, Eqs. (44) and (45) give $\mathbf{E}'_{\parallel} = 0$ and $\mathbf{E}'_{\perp} = 0$.

(c) Show that Eqs. (46) and (47) give $\mathbf{B}'_{\parallel} = 0$ and

$$\mathbf{B}'_{\perp} = \frac{\sqrt{c^2B^2 - E^2}}{cB}\mathbf{B}_{\perp} = \frac{\sqrt{c^2B^2 - E^2}}{cB}\mathbf{B}$$

(d) Since in the new reference frame there is a magnetic field but no electric field, the motion of a charged particle in this new reference frame is simply a circular orbit in a plane perpendicular to the magnetic field. As seen from the original reference frame, the circular motion around the magnetic field lines is supplemented by a translational velocity **V** perpendicular to the plane of **E** and **B**. Roughly sketch the resulting trajectory as seen in the original reference frame; do this for the case of positively and the case of negatively charged particles. The average translational velocity **V** perpendicular to the plane of **E** and **B** is called the $E \times B$ drift.

16. As described in the preceding problem, the equation of motion of a charged particle in uniform perpendicular electric and magnetic fields can be conveniently solved by performing a Lorentz transformation to a new reference frame moving perpendicularly to the plane of **E** and **B**. The preceding problem dealt with the case $E < cB$; here we deal with the case $E > cB$.

Under the assumption $E > cB$, take the velocity **V** of the Lorentz transformation as

$$\mathbf{V} = \frac{c^2(\mathbf{E} \times \mathbf{B})}{E^2}$$

(a) Show that $V < c$.
(b) Show that with this velocity **V**, Eqs. (46) and (47) give $\mathbf{B}'_{\parallel} = 0$ and $\mathbf{B}'_{\perp} = 0$.
(c) Show that Eqs. (44) and (45) give $\mathbf{E}'_{\parallel} = 0$ and

$$\mathbf{E}'_{\perp} = \frac{\sqrt{E^2 - c^2B^2}}{E}\mathbf{E}_{\perp} = \frac{\sqrt{E^2 - c^2B^2}}{E}\mathbf{E}$$

(d) Since in the new reference frame there is an electric field but no magnetic field, the motion of a charged particle is accelerated in the direction of the electric field. The resulting motion is "hyperbolic," of the type described in Section 4.6. (However, Section 4.6 deals only with a special case of one-dimensional motion, whereas now the particle may have an initial velocity in a direction different from that of the acceleration.) Roughly sketch the trajectory of the particle as seen in the original reference frame.

17. In an X-ray tube, a beam of high-speed electrons strikes a block of tungsten metal. The impact decelerates the electrons and thereby causes them to radiate intense electromagnetic radiation, or *Bremsstrahlung*. Consider an electron that has an initial energy of 20 keV, decelerates uniformly, and stops within a distance of 1.0×10^{-8} m. What is the magnitude of the electric radiation field at a distance of 0.20 m from the point of impact, in a direction perpendicular to the direction of acceleration? What is the magnitude of the magnetic radiation field?

18. From Eqs. (51) and (54), evaluate the Poynting vector $\mathbf{S} = \mathbf{E} \times \mathbf{B}/\mu_0 = c^2\epsilon_0\mathbf{E} \times \mathbf{B}$ and show that the power flowing in the direction θ is

$$S = \frac{1}{16\pi^2\epsilon_0 c^3}\frac{(qa\sin\theta)^2}{r^2}$$

Integrate this over the surface of a sphere of radius r, and show that the total power flowing out of this sphere is given by the **Larmor formula**

$$P = \frac{1}{4\pi\epsilon_0} \frac{2}{3} \frac{q^2a^2}{c^3}$$

Since this result is independent of the radius of the sphere, it indicates that the power continues flowing toward infinite distance, that is, the power is radiated away.

19. Verify that the vector sum of the horizontal electric field $\mathbf{E}_{\text{rad}}$ in Fig. 6.12a and the extra vertical electric field $\gamma\mathbf{V} \times \mathbf{B}_{\text{rad}}$ (with $B_{\text{rad}} = E_{\text{rad}}/c$) gives the tilted electric field shown in Fig. 6.12b, with the appropriate angle ϕ.

7
General Relativity

I WAS SITTING IN A CHAIR IN THE PATENT OFFICE AT BERN WHEN ALL OF A SUDDEN A THOUGHT OCCURRED TO ME: "IF A PERSON FALLS FREELY HE WILL NOT FEEL HIS OWN WEIGHT." I WAS STARTLED. THIS SIMPLE THOUGHT MADE A DEEP IMPRESSION ON ME. IT IMPELLED ME TOWARD A THEORY OF GRAVITATION.

Albert Einstein, Kyoto Lecture.

The theory of General Relativity originated from Einstein's attempts to formulate a relativistic theory of gravitation. The gravitational force acting on (and exerted by) an apple or a stone or some other body is proportional to its inertial mass. Because of this proportionality, the gravitational force acting on a body, or the weight, can be expressed as mg, where g is the local acceleration of free fall. In the equation of motion, $ma = mg$, the mass then cancels, and the acceleration of every kind of body is $a = g$, so the acceleration of free fall is universal, the same for every kind of body. However, according to Special Relativity, the mass depends on the energy stored in the body--the rest-mass energies of the particles in the apple contribute to the gravitational force, and so do the kinetic and potential energies of these particles, and so do thermal energies, and so do all other forms of energy. And Einstein found that in the context of Special Relativity, he could not construct a gravitational theory in which the gravitational force acting on every kind of body is exactly proportional to its mass.

But then it occurred to him that the universality of the acceleration of free fall can be regarded as an equivalence between acceleration and gravitation. Free fall of a reference frame located near some gravitating mass effectively eliminates gravitation in that reference frame, and an acceleration of a reference frame located in a region far from any gravitating mass creates artificial gravity in that reference frame. Einstein explored the consequences of this equivalence of acceleration and gravitation, and it ultimately led him to his theory of General Relativity.

We cannot deal with the complicated mathematical details of the theory of General Relativity in this book. Instead, we will concentrate on some of the simple consequences of the equivalence of acceleration and gravitation. We will see that this equivalence implies that clocks placed near a gravitating mass are subject to a new kind of time dilation, that is, they run slow relative to distant clocks placed outside of the gravitational field. Furthermore, we will see that light signals passing near a gravitating mass are deflected from the expected straight-line motion, and they are slowed down to a speed less than c. In Section 7.4 we will find that these effects lead to the inescapable conclusion that the spacetime near any gravitating mass is curved and that the ostensibly curved trajectories of particles or planets orbiting such a mass are merely the result of motion along the "straightest" possible path in this curved spacetime. And finally, in the last section we will briefly examine black holes, a spectacular instance of extreme spacetime curvature.

7.1 THE UNIVERSALITY OF THE ACCELERATION OF FREE FALL; THE PRINCIPLE OF EQUIVALENCE

It is a basic property of gravitation that at any given place, all bodies fall with the same acceleration. This property of gravitation was established by the experiments of Galileo late in the 16th century. Galileo checked not only that bodies of different masses fall with the same acceleration (9.80 m/s^2 at Pisa) but also that bodies made of different substances fall with the same acceleration. Since he did not have available precise clocks and other instruments for direct measurement of the rate of fall, he resorted to experiments with pendulums, and he compared the periods of oscillation of pendulums of equal lengths but with bobs made of different substances. Any inequality of the free-fall accelerations of the bobs would then show up as an inequality between their periods of oscillation. Galileo compared the periods of pendulum bobs of cork, lead, and other substances and found them equal. Such pendulum experiments were later repeated by Newton, with some refinements to compensate for the friction of air.

Experiments testing the universality of the acceleration of free fall of different substances have come to be called **Eötvös experiments** (pronounced ötvösh) after the Hungarian physicist Lorand von Eötvös, who around 1920 performed extensive tests of very high precision. For these experiments, Eötvös used a torsion balance, which is similar to a Cavendish balance (Fig. 7.1). Most later experiments used modified versions of such torsion balances; with these it is possible to test not only the downward rate of free fall toward the Earth but also the rate of free fall toward the Sun. The basic method underlying the experiments is most easily explained in the case of free fall toward the Sun, as follows: We attach the two masses that are to be compared to the beam of a torsion balance (see Fig. 7.2). The masses, the torsion balance, our laboratory, and the Earth are in free fall toward the Sun. If one of the masses is falling faster than the other, then the beam of the balance will tend to turn toward the Sun, that is, the balance will experience a torque. The balance will detect this torque and thereby reveal the faster rate of fall. If the torque is exactly zero, then the free-fall accelerations of the two masses are exactly equal.

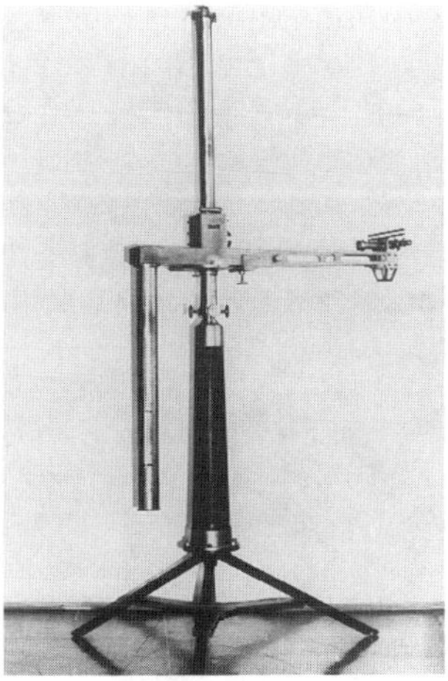

Fig. 7.1 One of the torsion balances used by Eötvös. (Geophysical Institute Eötvös Lorand, Budapest)

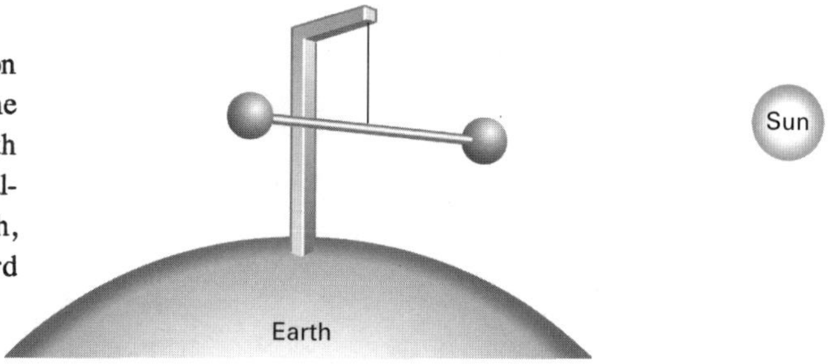

Fig. 7.2 A torsion balance located on the Earth, at the North Pole. The torsion balance, and the Earth, are in free fall toward the Sun.

In the hope of discovering some small deviation between accelerations of free fall, Eötvös and his successors tested a wide variety of different substances--many common metals as well as minerals and organic substances. The most precise experiments were performed by Dicke (at Princeton University) and Braginski (at Moscow University) on samples of gold, platinum, and aluminum. The atoms of gold and aluminum are about as different as they can be--one is very heavy, the other very light. Nevertheless, the delicate experiments of Dicke and Braginski show that these atoms fall with exactly the same acceleration, to within the experimental errors, about 1 part in 10^{13}. A satellite experiment developed by Everitt (Stanford University) and now awaiting launch in the Space Shuttle is expected to test this equality to even higher precision, about 1 part in 10^{17}.

The available experimental evidence confirms what Galileo recognized four centuries ago: under the influence of gravity, all kinds of bodies fall with the same acceleration. One consequence of this universality of the acceleration is that free fall simulates the condition of weightlessness, or absence of gravity. For example, Fig. 7.3 is a photograph of an astronaut floating freely above the cargo bay of the Space Shuttle in orbit around the Earth. The Space Shuttle and the astronaut are in free fall. The astronaut feels weightless; he is in a zero-*g* environment. The reason for this apparent absence of gravitational forces is that any object released by the astronaut, and even his dangling arm or leg, falls with exactly the same acceleration as the astronaut; that is, relative to the astronaut, it seems not to fall at all.

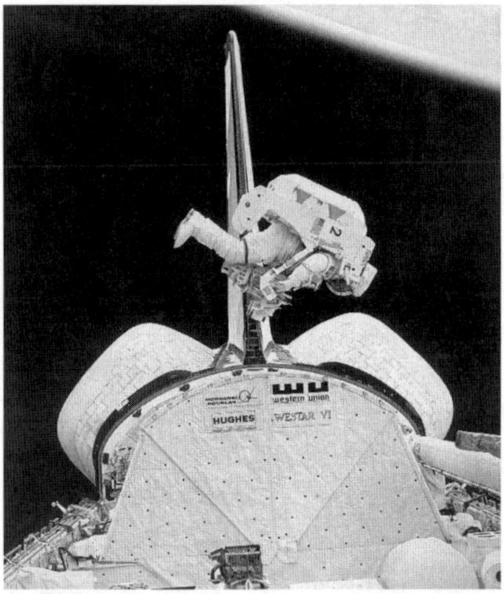

Fig. 7.3 Astronaut floating freely above the cargo bay of the Space Shuttle. (NASA)

The freely falling reference frame of the Space Shuttle simulates an inertial reference frame: a body at rest remains at rest, and a body in motion continues to move along a straight line. Such a reference frame in free fall under the influence of gravitation is called a **local inertial reference frame**. This kind of reference frame must be restricted to a reasonably small region--it can simulate a true inertial reference frame only over a small range of distances. For instance, if we attempt to extend the reference frame of the Space Shuttle over a large range of radial distances, then the decrease of the gravitational acceleration with distance from the Earth would cause a free body in the upper part of the reference frame to accelerate less than a body in the lower part--both of these bodies would then have an acceleration relative to the midpoint of the reference frame (see Fig. 7.4). Although these deviations between the accelerations of the bodies in a freely falling reference frame are most noticeable if the reference frame is large, a sensitive instrument can detect the deviations even if the distances are small. Hence the elimination of gravity by free fall is not perfect, but the residual effects are usually so small that we can ignore them.

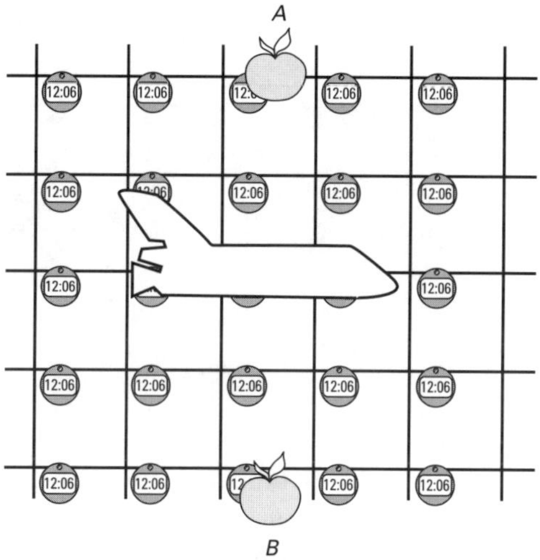

Fig. 7.4 In a large reference frame attached to the Space Shuttle, a body released at the top (*A*) has a smaller acceleration than a body released at the bottom (*B*). Relative to the center of the Space Shuttle, both of these bodies move away.

The high-precision experiments on the equality of the accelerations of free fall establish that as far as mechanical experiments are concerned, there is no intrinsic distinction between a local inertial reference frame in a gravitational field and a true inertial reference frame in interstellar space, far from any gravitating masses. Furthermore, the cancellation of the effects of gravitation and the effects of acceleration also means that there is no intrinsic distinction between an accelerated, noninertial reference frame in some region far from any gravitating masses and a reference frame at rest in a gravitational field. The acceleration of the noninertial reference frame mimics a gravitational field, that is, the acceleration generates extra artificial gravity. We experience this kind of artificial gravity whenever we ride an elevator accelerating upward; in this case we experience some artificial gravity (extra weight) added on to the normal gravity. Astronauts in training are subjected to large artificial gravity, up to $20g$, generated in centrifuges (see Fig. 7.5). And even larger values, up to $80g$, were endured briefly by a volunteer in a rocket sled.

The lack of an intrinsic distinction between the effects of gravity and the effects of acceleration suggested to Einstein that there is not only a relativity of uniform translational motion but also a relativity of acceleration--astronauts in a spacecraft in free fall in a gravitational field can detect neither their acceleration nor the gravitational field by any mechanical experiment performed within or near their spacecraft,

Fig. 7.5 Large centrifuge used to generate artificial gravity. The astronauts ride in the gondola at the left. (NASA)

and astronauts in an accelerated spacecraft far from any gravitating mass cannot tell whether they are experiencing an acceleration or a genuine gravitational field. Einstein conjectured that, as in the case of the Principle of Special Relativity, what is true for mechanical experiments should also be true for any other kind of experiment. He stated this conjecture as his **Principle of Equivalence,** a hypothesis about the character of gravitation:

In a local inertial reference frame the effects of gravity are absent; in such a reference frame, all the laws of physics are the same as in a true inertial reference frame. Conversely, in an accelerating reference frame far from any gravitating masses there is an apparent gravitational field, and all the laws of physics are the same as though the reference frame were at rest near a gravitating body.

We might call this the Principle of General Relativity, since it asserts a relativity of acceleration that goes beyond the relativity of velocity asserted by the Principle of Special Relativity. But the relativity of acceleration is severely limited--it applies only to local inertial reference frames, not to all reference frames. In contrast to the Principle of Special Relativity, which imposes restrictions on all the laws of physics, the Principle of Equivalence imposes restrictions only on the laws of gravitation. It is a hypothesis about the basic properties of gravitation.

Einstein published the Principle of Equivalence in 1911. He explored the consequences of this principle for several years, and it finally led him to his theory of General Relativity. In the next sections we will see that this principle has surprising consequences for the behavior of clocks and of light signals in a gravitational field.

7.2 *THE GRAVITATIONAL TIME DILATION*

We know from the theory of Special Relativity that whenever a clock is moving at high speed relative to another clock, it runs slows, that is, it suffers a time dilation. As we saw in Section 3.1, this kind of time dilation is a direct consequence of the Principle of Special Relativity. We will now see that the Principle of Equivalence implies that there is an extra kind of time dilation, called the **gravitational time dilation**, that occurs whenever a clock placed in a low gravitational potential is compared with another clock placed at a higher potential, that is, whenever a clock placed near a gravitating mass is compared with another clock placed farther away.

Suppose that one of the clocks is at rest at a position r_1 near the gravitating mass, and the other clock is at rest at a position r_2 farther away (see Fig. 7.6). We assume that the clocks are of identical construction so that, in the absence of relativistic effects, they would tick at the same rate. In practice the "clocks" might be two identical oscillating atoms, which guarantees their identical construction. Also, we assume that the clocks are not directly affected by the gravitational field (in contrast to, say, pendulum clocks, whose rates are affected by the gravitational force in a very direct way).

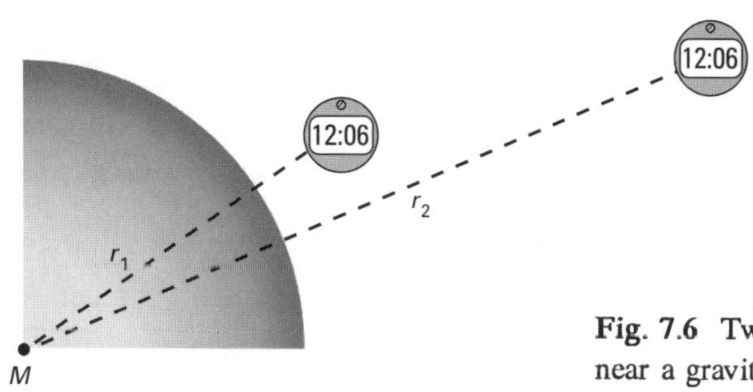

Fig. 7.6 Two clocks at rest near a gravitating mass.

To compare the rates of these clocks, we take a small reference frame and let it fall freely from r_2 and r_1. This reference frame is a local inertial reference frame, and according to the Principle of Equivalence, its clocks behave like the clocks of an inertial reference frame, that is, they run at a fixed rate, and they can serve as standards of time. We can therefore use the freely falling reference frame as a traveling clock that permits us to compare clocks 1 and 2 in two steps: When the freely falling reference frame is at r_2, we compare the rate of clock 2 with the rate of the clocks of the reference frame; and when the reference frame reaches r_1, we compare the rate of clock 1 with the rate of the clocks of the reference frame. Suppose that at r_2 the freely falling reference has a speed v_2, and at r_1 a speed v_1. Since we want to make the comparisons of the clock rates "on the fly," without stopping the freely falling reference frame, we must take into account the special-relativistic time dilation. When the freely falling reference frame passes by r_2, a proper time interval $d\tau_2$ registered by clock 2 is measured by the clocks of the freely falling reference frame as a dilated time interval $dt_2 = d\tau_2/\sqrt{1 - v_2^2/c^2}$. Likewise, when the freely falling reference frame reaches r_1, a proper time interval $d\tau_1$ registered by clock 1 is measured as a dilated time interval $dt_1 = d\tau_1/\sqrt{1 - v_1^2/c^2}$. In a fixed time interval $dt_2 = dt_1$ reckoned in the freely falling reference frame, the clocks 1 and 2 therefore register different proper time intervals $d\tau_1$ and $d\tau_2$ such that

$$\frac{d\tau_2}{\sqrt{1 - v_2^2/c^2}} = \frac{d\tau_1}{\sqrt{1 - v_1^2/c^2}} \tag{1}$$

or

$$\frac{d\tau_2}{d\tau_1} = \frac{\sqrt{1 - v_2^2/c^2}}{\sqrt{1 - v_1^2/c^2}} \tag{2}$$

If the point r_1 is at a lower height than the point r_2, then $v_1 > v_2$, and therefore $d\tau_2 > d\tau_1$. Thus, the clock at r_1 shows a smaller time interval, that is, it runs slow compared with clock 2.

We can express Eq. (2) in terms of the gravitational potential, for which purpose we approximate the square roots in the usual way:

$$\frac{d\tau_2}{d\tau_1} = 1 - \frac{v_2^2}{2c^2} + \frac{v_1^2}{2c^2} \tag{3}$$

During the free-fall motion in a gravitational field, the sum of the kinetic and gravitational potential energies is constant. Introducing the gravitational potential Φ, we can express this conservation of energy as $mv^2/2 + m\Phi = \text{const}$. Here the gravitational potential Φ is defined as the gravitational potential energy per unit mass. For instance, for the approximately constant gravitational force near the surface of the Earth, $\Phi = gh$. But for the inverse-square gravitational force produced by a spherical mass, $\Phi = -GM/r$ (here the negative sign in the potential has been chosen by analogy with the Coulomb potential for an attractive electric force.) Accordingly, the change in $v^2/2$ equals the negative of the change in Φ, and Eq. (3) becomes

$$\boxed{\frac{d\tau_2}{d\tau_1} = 1 + \frac{\Phi_2}{c^2} - \frac{\Phi_1}{c^2}} \tag{4}$$

This says that if $\Phi_2 > \Phi_1$, then $d\tau_2 > d\tau_1$, which confirms that the clock at the lower height, and the lower gravitational potential, runs slow.

We can apply Eq. (4) to clocks on or near the surface of the Earth. For instance, if clock 1 is on the surface of the Earth, and clock 2 is at some modest height h above the surface, then $\Phi_2 - \Phi_1 \simeq gh$, and

$$\frac{d\tau_2}{d\tau_1} \simeq 1 + \frac{gh}{c^2} \tag{5}$$

In practice, the simplest experimental method for the comparison of clocks uses periodic signals sent from one clock to the other, such as light signals in the form of a periodic light wave, where each wave-crest emitted by the clock (or by an atom) counts as a tick. The time dilation of a clock then leads to an apparent frequency shift for the light wave. If the clock runs slower than normal, the frequency of the light wave measured with this clock will appear to be higher than normal. Thus, the ratio ν_1/ν_2 of the frequencies measured with clocks 1 and 2 is inversely proportional to $d\tau_2/d\tau_1$. Accordingly, Eq. (5) gives

$$\frac{\nu_1}{\nu_2} \simeq 1 + \frac{gh}{c^2} \tag{6}$$

from which we see that the fractional frequency shift is

$$\boxed{\frac{\nu_2 - \nu_1}{\nu_2} \simeq -\frac{gh}{c^2}} \tag{7}$$

This apparent decrease of frequency for a light wave emitted at low altitude and received at a higher altitude in a gravitational potential is called the **gravitational redshift.**

Einstein first deduced the gravitational redshift from a very simple Doppler-shift argument, also based on the Principle of Equivalence. He replaced the two stationary clocks located in the gravitational field of the Earth (Fig. 7.7a) by two clocks located in an accelerated spacecraft in interstellar space, far from any gravitating mass

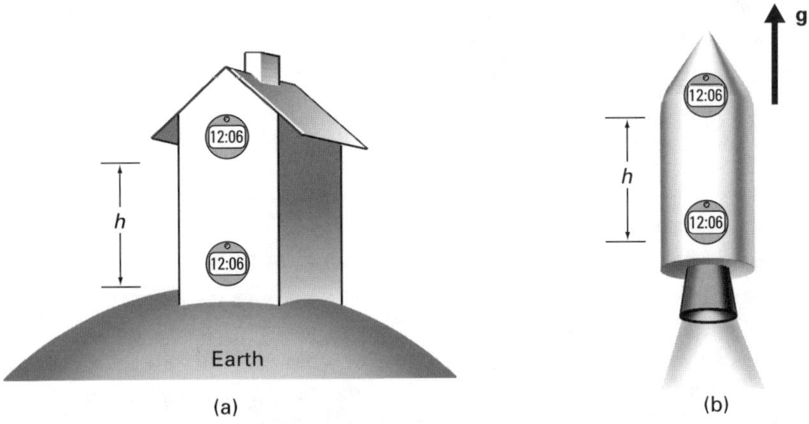

Fig. 7.7 (a) Two clocks at rest in the gravitational field of the Earth.
(b) Two clocks located in an accelerated spacecraft. One clock is at the bottom of the spacecraft, and one is at the top, at a distance h from the other clock.

(Fig. 7.7b). According to the Principle of Equivalence, if the accelerated spacecraft has an acceleration g, then the behavior of its clocks will mimic the behavior of the clocks in the gravitational field of the Earth. Einstein could then deduce the redshift by examining what happens to a light signal sent from the lower clock to the upper clock in the accelerated spacecraft. If the distance between the clocks is h, then the time for a light signal to travel from the lower clock to the upper clock is approximately h/c, and in this time interval the acceleration g gives the upper clock a velocity increment $\Delta v \simeq g \times h/c$, which is added to whatever velocity both clocks had at the time of emission. Because of this recessional velocity of the upper clock relative to the point of emission, the light signal arrives at the upper clock with a Doppler shift. According to the familiar Doppler-shift formula, the fractional frequency shift is

$$\frac{\nu_2 - \nu_1}{\nu_2} = -\frac{\Delta v}{c} \simeq -\frac{gh}{c^2} \tag{8}$$

This coincides with the gravitational redshift (7) we found above.

■ **EXAMPLE 1.** What is the value of the gravitational redshift for $h = 20$ m? for $h = 10$ km?

SOLUTION. For $h = 20$ m, Eq. (7) gives a fractional frequency shift

$$\frac{\nu_2 - \nu_1}{\nu_2} = -\frac{gh}{c^2} = -\frac{9.8 \text{ m/s}^2 \times 20 \text{ m}}{(3.0 \times 10^8 \text{ m/s})^2} = -2.2 \times 10^{-15}$$

For $h = 10$ km, it gives

$$\frac{\nu_2 - \nu_1}{\nu_2} = -\frac{gh}{c^2} = -\frac{9.8 \text{ m/s}^2 \times 10^4 \text{ m}}{(3.0 \times 10^8 \text{ m/s})^2} = -1.1 \times 10^{-12}$$

Gravitational redshifts were measured by astronomers in the light emitted by atoms on the surface of the Sun and on the surface of other stars. These early measurements established that a redshift exists, but they were afflicted with large experimental uncertainties [2]. The first measurement of reasonable precision was performed by Pound and Rebka [3] at Harvard University who used gamma rays "climbing" upward in the gravitational potential of the Earth over a height of about 20 m. As we saw in Example 1, for such a height, the fractional frequency shift is quite small, only about 2 parts in 10^{15}. The gamma rays were emitted by a source consisting of nuclei of iron (^{57}Fe) located at ground level, and they were aimed toward an absorber consisting of similar nuclei placed at a height of 22 m (see Fig. 7.8). Measurement of the small frequency shift was possible with this arrangement because the resonant absorption of gamma rays by nuclei of iron is very sensitive to the frequency of the rays. The experiment relied on the Mössbauer effect (see Section 5.5) to prevent loss of any of the gamma-ray energy by recoil during emission or absorption. The best version of this experiment [4] confirmed the redshift formula (8) to within an experimental error of about 1%.

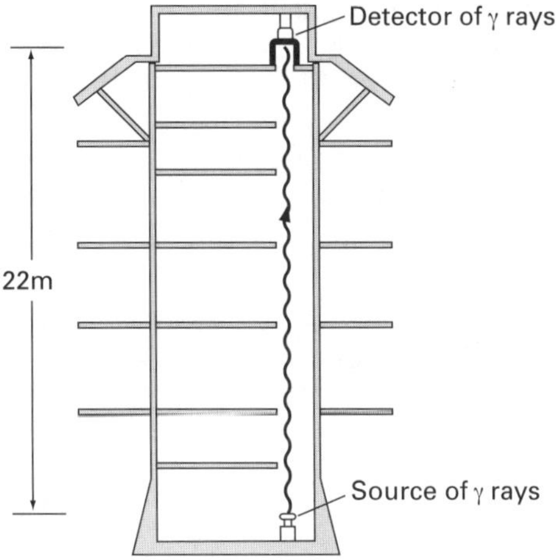

Fig. 7.8 Absorber and emitter of gamma rays for the time-dilation experiment of Pound and Rebka.

A much more precise experiment was performed by Vessot et al. [5], who used a Scout rocket to lift an atomic clock to an altitude of 10,000 km above the surface of the Earth (Fig. 7.9). For this height, the redshift of an upward signal (or the blueshift of a downward signal) is about 4×10^{-10}. As already described in Section 3.3, the onboard clock was also affected by the special-relativistic time dilation, except at apogee, when the speed of the rocket is instantaneously zero. During the flight of the rocket, radio signals sent both ways between

Fig. 7.9 Atomic maser clock (in dome) before its installation in the nose cone of the Scout rocket. (R. F. C. Vessot, Harvard-Smithsonian Center for Astrophysics)

the on-board clock and a clock on the ground permitted comparison of the rates of these clocks. The experimental error of this experiment was about 0.02%

The gravitational time dilation was also confirmed by Häfele and Keating [6] and by Alley [7], who used aircraft to lift atomic clocks to about 10 km, an altitude routinely attained by commercial jet aircraft. In the first of these experiments [6], the atomic clocks were carried in the aircraft during lengthy trips on regularly scheduled flights around the world; in the second experiment [7], the atomic clocks were carried in an aircraft that flew in a prolonged high-altitude holding pattern over an airfield. The clocks were affected by the special-relativistic time dilation (see Section 3.3), but they were also affected by the gravitational time dilation because of their altitude. According to Eq. (5), clocks carried to a height of 10 km run faster than clocks on the ground by about 1 part in 10^{12}, and this was tested, and confirmed, by bringing the clocks back to their starting point and comparing them with clocks left on the ground. Although these tests did not achieve the high precision of the rocket experiment, they had the advantage of showing directly and conclusively that clocks in a gravitational field run slow and that the redshift must be attributed to this slowing of the clocks (or slowing of the natural vibration of the atoms), not to some kind of frequency loss or energy loss of the light or radio wave as it "climbs" upward in the gravitational field.

7.3 THE BENDING OF RAYS OF LIGHT

When Einstein discovered the gravitational time dilation, he immediately recognized that this dilation implies that the speed of light near a gravitating mass must be slower than the speed of light far from the mass. The easiest way to recognize this is to recall that we can construct a clock by bouncing a light signal back and forth between two mirrors separated by a fixed distance (the "light clock" discussed in Section 3.1). If such a clock suffers a time dilation in the space near a gravitating mass, it must be because the light signal has a reduced speed. This means that the space near a gravitating mass is like a refractive medium, with a higher index of refraction near the mass. This kind of refractive medium will deflect, or bend, rays of light, according to the usual rule that the rays bend toward the region where the index of refraction is largest. Thus, a ray of light passing by a large mass, such as that of the Sun, will be deflected toward the mass.

Such a deflection had already been contemplated by P. de Laplace, two centuries ago. At that time it was thought that light consists of particles, and in that case the trajectory of a light ray near the Sun would be expected to take the shape of a hyperbola (according to Newtonian physics, this is the trajectory for a particle of positive energy in a gravitational field). When the particle theory of light was abandoned in favor of a wave theory early in the 19th century, speculations about the deflection were also abandoned until Einstein reinvented the deflection in the context of his theory of General Relativity.

The deflection of light can also be understood directly from the Principle of Equivalence. Consider a spacecraft in free fall toward the Sun. The spacecraft is a local inertial reference frame, and gravity is effectively absent in it. A light ray will propagate across the spacecraft in a straight path, just as in interstellar space. But if the light ray follows a straight path relative to the spacecraft, then it must follow a curved path relative to the Sun. Figure 7.10 shows the light ray propagating from the left side of the spacecraft to the right side. The light ray starts at the midpoint of the left side and ends at the midpoint of the right side. Since the spacecraft accelerates downward, the light ray must also accelerate downward at the same rate if it is to follow a straight path across the spaceship. Thus, the Principle of Equivalence leads us to the qualitative conclusion that in the gravitational field of the Sun, a light ray must follow a curved path--the gravitational field deflects rays of light.

For a quantitative calculation of the deflection, we can exploit the formula for the gravitational time dilation. From this it is easy to deduce a quantitative expression for the slowing of the speed of light, and from this, in turn, it is possible to calculate the deflection. We will

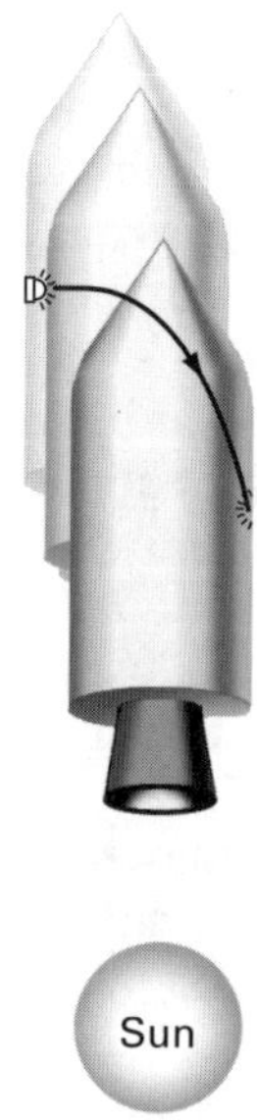

Fig. 7.10 A light ray propagating across a spacecraft accelerating downward, toward the Sun.

not bother with this calculation here, because the result obtained by these means is not quite right. The trouble is that besides the deflection caused by the slowing of light, there is an extra deflection caused by a distortion of the geometry of space. In his first calculation of the deflection, Einstein obtained the wrong result because he was not aware of this second contribution; he corrected the calculation later, after he completed his formulation of the theory of General Relativity.

The corrected value of the deflection is 1.75 arcseconds for a light ray that passes close to the edge of the Sun (a grazing light ray; see Fig. 7.11). If the ray of light originates from a distant star, then an astronomer on Earth will see this star displaced from its usual position whenever the Sun happens to be near the line of sight (see Fig. 7.12). But because the deflection is so small, it is very difficult to detect. An extra complication is that the astronomer on the Earth must observe the star when it is near the Sun in the sky, and this is possible only during an eclipse of the Sun, when the Sun's glare is blocked by the Moon. The first measurements of the deflection were made during the eclipse of 1919, and the observational results, which were in reasonable agreement with Einstein's prediction, created a sensation. Since then, the observations have been repeated during many other eclipses. Deflections were detected, but it has proved impossible to achieve good precision.

Rather more reliable measurements of the deflection have been made with radio waves. With their radiotelescopes, radioastronomers discovered many quasars in the sky. These are pointlike sources of radio waves, well suited to deflection measurements. One of these quasars (3C273) is near the Sun's path across the sky, and when the

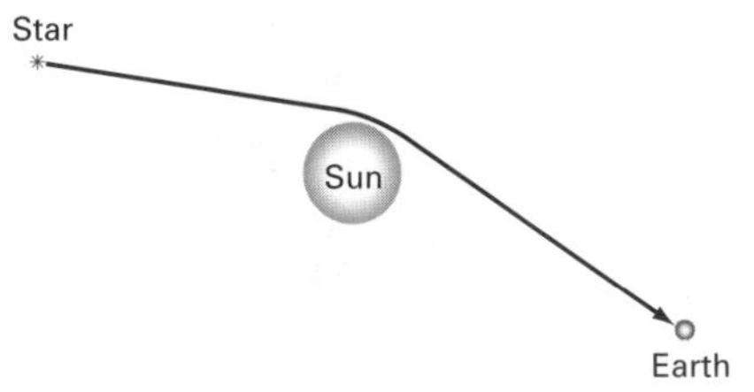

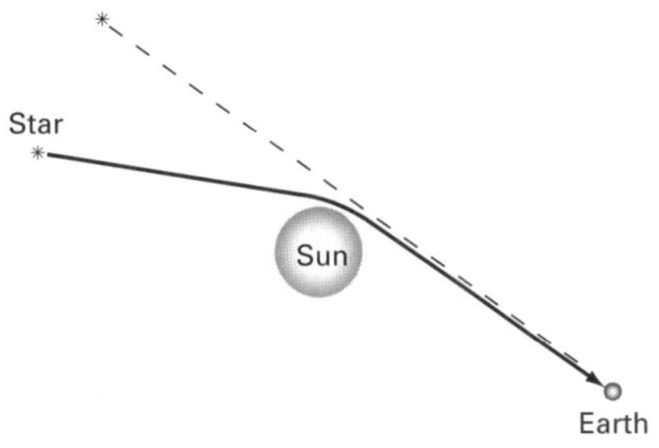

Fig. 7.11 (Left) A grazing light ray from a star passes close to the edge of the Sun.

Fig. 7.12 (Right) Apparent displacement of the star as seen from the Earth.

Sun is near our line of sight to this quasar, the radio waves are deflected in the same way as light rays. Measurements of this deflection of radio waves have been made with several of the large radiotelescopes, and these measurements have given us a precise test of Einstein's prediction, to within the observational error of about 0.01%. [2].

Besides measuring the deflection of radio waves caused by the Sun, radio astronomers have observed deflections caused by distant galaxies. When a distant galaxy is near or on the line of sight from an even more distant galaxy, the deflection of the path of the radio waves can result in spectacular distortions of the image of the more distant galaxy. For instance, if the more distant galaxy is almost exactly on the line of sight, its image as seen by the radiotelescopes on the Earth is distorted into a ring, called an **Einstein ring** (see Fig. 7.13). This ring results because the radio waves passing around the intervening galaxy are deflected from all sides toward the central line. The intervening galaxy is called a **gravitational lens**. But in contrast to a genuine lens, the gravitational lens produces severely distorted images and sometimes multiple images, and it has no well-defined focal point. Radio astronomers have discovered many such cases of galaxies acting as gravitational lenses, and they have even discovered some cases of whole clusters of galaxies acting jointly as gravitational lenses.

The slowing of radio waves in a gravitational field has been confirmed directly by experiments with radar signals sent past the Sun. Shapiro et al. [8] and other radioastronomers [2] have used their large radiotelescopes as transmitters to send radar signals to Mercury and to Venus, from where the signals are reflected back to Earth. If a planet is "in opposition," so the Sun is near our line of sight to the planet, then the radar signals are slowed down, and the radar echo suffers a retardation, that is, an extra delay over and above the normally expected travel time. The maximum retardation is about 10^{-4} s for a grazing radar signal. Several measurements of this retardation have confirmed it to within a few percent.

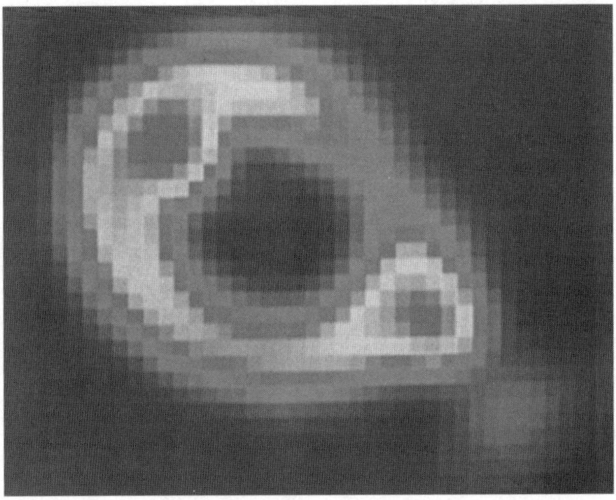

Fig. 7.13 The Einstein ring MG1131+0456 as observed with a radiotelescope. The intervening galaxy that acts as gravitational lens is not visible in this radio image. (J. N. Hewitt, Massachusetts Institute of Technology)

7.4 CURVED SPACETIME; THE THEORY OF GENERAL RELATIVITY

In the preceding section we saw that the Principle of Equivalence implies that the speed of light is slowed down in a gravitational field, and we reviewed the experimental evidence confirming this. But the Principle of Equivalence also implies that the speed of light ought to be the same everywhere in a gravitational field, which contradicts our previous conclusion. Before we explore the consequences of this contradiction, let us see how the Principle of Equivalence implies a constant speed of light.

Imagine that everywhere around the Sun there are small spacecraft (Space Shuttles), instantaneously at rest but in free fall; that is, each spacecraft has just been released, and it is accelerating toward the Sun but has not yet acquired any significant velocity. Each spacecraft is a local inertial reference frame, and in each, gravity is effectively absent. When astronauts aboard each spacecraft measure the speed of light in their immediate vicinity, the Principle of Equivalence demands that they all obtain the same result as in interstellar space, far from the Sun, that is, they all obtain 3.0×10^8 m/s.

We now have a conflict between this new determination of the speed of light and the previous determination described in the preceding section. Our new determination is done directly, on a small scale, in local inertial reference frames. We will call this the *local* determination. The previous determination is done indirectly, on a large scale, by timing the light signal over a long distance (as in the retardation experiment) or by using remote instruments to deduce the speed (as in the deduction of the speed of light from the gravitational time dilation). We will call this the *global* determination.

To grasp the implications of the conflict between the local speed of light and the global speed of light, the following simple analogy is helpful. Imagine two towns, one exactly west of the other, separated by a horizontal distance, or a "chart" distance measured naively on a surveyor's chart, of exactly 100 km. A highway connecting the two towns runs straight west. Imagine that we send a car from one town to the other along this highway at a speed, indicated by the speedometer, of exactly 100 km/h. We then expect that the car takes exactly 1 hour to complete the trip. But suppose that when we performed the experiment, we found that the car takes more than 1 hour, which would yield a calculated speed of less than 100 km/h. What would we conclude from this discrepancy? We would conclude that the highway climbs over some hills or dips into some valleys between the towns, and that the actual distance measured along the highway is larger than the horizontal distance, or "chart" distance. That is, we would conclude that the space between the townships is not a flat surface but a curved surface. This is the only possible explanation of the conflict between the local determination of the speed (as indicated by the speedometer) and the global determination of the speed (as calculated from the "chart" distance and the travel time).

The explanation of the conflict between the locally and the globally determined speeds of light in the space surrounding a gravitating mass is similar. The globally determined speed is less than the locally determined speed because the space is curved, and the actual distance along the path of a light ray is longer than what we calculate by assuming a flat geometry of space. But the analogy between the speed of a car and the speed of a light signal is not perfect. For the light signal traveling through the space surrounding a gravitating mass, there is an additional complication in that time is also "curved," or, more precisely, spacetime is curved. We can recognize the distortion of time from the gravitational time dilation. If time were the undistorted time of Special Relativity, with no strange features except for the usual special-relativistic time dilation, then a clock transported very slowly into a gravitational field and then brought back would display no retardation relative to a stationary clock (we know from Section 3.1 that for the limiting case of a transport speed that approaches zero, the special-relativistic time dilation approaches zero). Hence the retardation displayed by a clock transported very slowly into a gravitational field and then brought back indicates a distorted time, that is, a "curved" time.

The fundamental tenet of Einsteins' theory of General Relativity is that gravity is caused by a curvature of spacetime. The Sun, or any other mass, produces a curvature in the spacetime surrounding it, and this curvature affects the motion of light and the motion of planets. The Sun does not exert any force on the light or the planets. Rather, it distorts the geometry of spacetime, and the light or the planets merely move on the "straightest" possible paths in this distorted spacetime geometry. The curved paths that we observe light and planets to follow merely represent the "straightest" possible motion in a curved spacetime geometry. Thus, Einstein's theory of General Relativity tells us that gravity is merely a manifestation of geometry.

We are not able to visualize the curved four-dimensional spacetime of General Relativity. We cannot even visualize a curved three-dimensional space. At most, we can visualize a curved two-dimensional space, that is, a curved surface that bends and curves into a third dimension. Hence, to visualize the curved spacetime in the vicinity of the Sun, we will have to restrict ourselves to two of its four dimensions. One way to do this is to consider only those points that are in the equatorial plane of the Sun (see Fig. 7.14). The curved geometry of this plane can be displayed by deforming this plane into a third dimension. Figure 7.15 shows the deformed plane; in this figure, the third dimension (up and down) merely provides the means to display how the geometry of the equatorial plane is curved. When we move along the deformed surface in Fig. 7.15, we are not actually moving up or down in real space; we are always staying in the equatorial plane of the Sun.

The curved surface in Fig. 7.15 has the shape of a bowl, and distances measured along the lines that "dip" into the bowl are longer than they would be on a flat surface. Obviously, the geometry on the curved surface deviates from Euclidean geometry. For instance, the circumference of a concentric circle is less than 2π times the radial distance from the center measured along the bowl. For a circle along the Sun's equator, this deficit of circumference is 23 km; for the Earth, it is only 5.9 mm. These deviations from Euclidean geometry are too small for any direct measurement. The curvature of space reveals itself only indirectly by its effects on the global speed of light and on the motion of the planets.

Mathematically, the curving of spacetime is expressed by a modification of the spacetime interval. We know from Chapter 2 [see Eq. (2.29)] that the flat spacetime of Special Relativity is characterized by a spacetime interval $(\Delta s)^2 = c^2(\Delta t)^2 - (\Delta x)^2 - (\Delta y)^2 - (\Delta z)^2$, or, written in terms of differentials,

$$(ds)^2 = c^2(dt)^2 - (dx)^2 - (dy)^2 - (dz)^2 \tag{10}$$

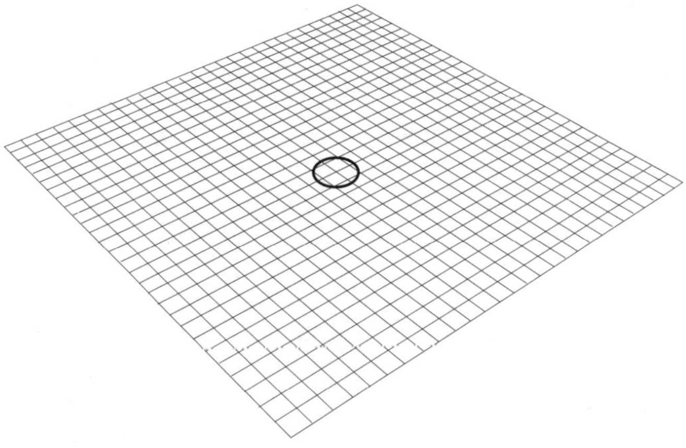

Fig. 7.14 The equatorial plane of the Sun.

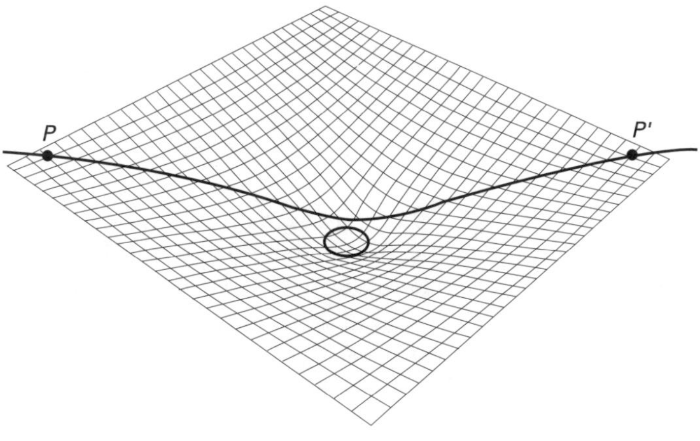

Fig. 7.15 The curved geometry in the equatorial plane can be visualized as a deformation of the plane. The straight lines, or straightest lines, in the curved geometry have to follow the curving surface.

In a curved spacetime, this simple expression for the spacetime interval is modified. Each term on the right side is multiplied by a coefficient which is a function of position:

$$(ds)^2 = g_{00}c^2(dt)^2 + g_{11}(dx)^2 + g_{22}(dy)^2 + g_{33}(dz)^2 \qquad (11)$$

Here the coefficients g_{00}, g_{11}, g_{22}, and g_{33} are written in a somewhat complicated form, with two subscripts each, because they are components of a tensor quantity, the **metric tensor**. In general, this tensor also has off-diagonal components g_{01}, g_{12}, etc., and the right side of

Eq. (11) has extra terms $g_{01}cdxdt$, $g_{12}dxdy$, etc. But for a time-independent, static spacetime, the spacetime interval can always be written in terms of purely diagonal terms, as in Eq. (11), and we will here deal only with this simple case. Note that the spacetime interval of Special Relativity is evidently a special case of Eq. (11), with $g_{00} = 1$ and $g_{11} = g_{22} = g_{33} = -1$.

According to General Relativity, the spacetime in the vicinity of any gravitating mass is characterized by a modified spacetime interval with coefficients g_{00}, g_{11}, g_{22}, and g_{33} different from ± 1. The coefficients g_{00}, g_{11}, g_{22}, and g_{33} depend on the distribution of mass, or, more precisely, on the distribution of energy and momentum. Einstein formulated a set of equations from which these coefficients can be calculated. These equations are complicated tensor equations, but for the simple case of a spherical mass M, such as the Sun or a planet, the coefficients calculated from the Einstein equations are quite simple. The coefficients are functions of the radial distance, and they are approximately

$$g_{00} = 1 - \frac{2GM}{rc^2} \tag{12}$$

$$g_{11} = g_{22} = g_{33} \simeq -1 - \frac{2GM}{rc^2} \tag{13}$$

These coefficients are called the **Schwarzschild solution.**

We can verify that with this value of the coefficient g_{00}, the curved spacetime interval (11) gives us the expected gravitational time dilation. Consider a clock at rest at a radial position r. For this clock $dx = dy = dz = 0$, and hence, according to Eq. (11), the proper time interval between the ticks of the clock is

$$(d\tau)^2 = \frac{(ds)^2}{c^2} = g_{00}(dt)^2 = \left(1 - \frac{2GM}{rc^2}\right)(dt)^2 \tag{14}$$

Therefore,

$$d\tau = \sqrt{1 - \frac{2GM}{rc^2}}\, dt \simeq \left(1 - \frac{GM}{rc^2}\right) dt \tag{15}$$

Since $-GM/rc^2$ equals the Newtonian potential Φ, Eq. (15) can be expressed as

$$d\tau = (1 + \Phi/c^2)dt \tag{16}$$

This agrees with our previous result (4) for the gravitational time dilation of clocks. In fact, Eq. (16) is a special case of Eq. (4), with $d\tau_2 = d\tau$, $\Phi_2 = \Phi$ and with $d\tau_1 = dt$, $\Phi_1 = 0$ (as is appropriate if clock 1 is far from the mass M, where the spacetime is flat).

From Eq. (11), we can also easily obtain a formula for the speed of light in the space surrounding a spherical mass. In General Relativity, as in Special Relativity, a light signal is characterized by a zero spacetime interval. Thus,

$$0 = g_{00}c^2(dt)^2 + g_{11}(dx)^2 + g_{22}(dy)^2 + g_{33}(dz)^2 \tag{17}$$

With Eqs. (12) and (13), this becomes

$$0 = \left(1 - \frac{2GM}{rc^2}\right)(dt)^2 - \left(1 + \frac{2GM}{rc^2}\right)[(dx)^2 + (dy)^2 + (dz)^2] \tag{18}$$

Accordingly, the speed of light is*

$$\frac{dl}{dt} = \frac{\sqrt{(dx)^2 + (dy)^2 + (dz)^2}}{dt} = c \times \frac{\sqrt{1 - 2GM/rc^2}}{\sqrt{1 + 2GM/rc^2}} \tag{19}$$

The factor $\sqrt{1 - 2GM/rc^2}$ in the numerator on the right side results from the gravitational time dilation, and the factor $\sqrt{1 + 2GM/rc^2}$ in the denominator results from the curvature of space. Both of these factors lead to a reduction of the speed of light. Note that this reduced speed of light is the "global" speed; it is the speed calculated according to the time t and the distances dx, dy, dz determined by remote measuring instruments. In the language of our analogy with the car traveling on the highway, the differentials dx, dy, dz are "chart" distances, not true distances. The reduction of the speed by the factor $\sqrt{1 + 2GM/rc^2}$ results from the discrepancy between the "chart" distances dx, dy, dz and the true distance. In the vicinity of a body of large mass, the speed of light is reduced by a significant amount.

■ **EXAMPLE 2.** By what factor is the speed of light reduced near the surface of the Earth? By what factor is the speed of light reduced near the surface of a neutron star, with a mass of 2×10^{30} kg and a radius of 10 km?

SOLUTION. For the surface of the Earth, with $M = 6.0 \times 10^{24}$ kg and $r = 6.4 \times 10^6$ m, the factor by which the speed of light is reduced is

$$\frac{\sqrt{1 - 2GM/rc^2}}{\sqrt{1 + 2GM/rc^2}} \simeq \sqrt{1 - 2GM/rc^2} \times \sqrt{1 - 2GM/rc^2}$$

$$\simeq 1 - \frac{2GM}{rc^2} = 1 - 2 \times \frac{6.67 \times 10^{-11}\ \text{N·m}^2/\text{kg}^2 \times 6.0 \times 10^{24}\ \text{kg}}{6.4 \times 10^6\ \text{m} \times (3.0 \times 10^8\ \text{m/s})^2} = 1 - 1.4 \times 10^{-9}$$

* As stated above, Eqs. (13) are only approximations. Correspondingly, Eq. (19) is an approximation. The exact result for the speed of light is $dl/dt = c \times \sqrt{1 - 2GM/rc^2} \times \sqrt{1 - 2GM/rc^2}$, which is approximately the same as Eq. (19).

For the neutron star, with $M = 2 \times 10^{30}$ kg and $r = 10^4$ m, the factor is

$$\simeq 1 - \frac{2GM}{rc^2} = 1 - 2 \times \frac{6.67 \times 10^{-11}\ \mathrm{N{\cdot}m^2/kg^2} \times 2 \times 10^{30}\ \mathrm{kg}}{10^4\ \mathrm{m} \times (3.0 \times 10^8\ \mathrm{m/s})^2} = 0.3$$

Thus, on the Earth, the reduction of the speed of light is insignificant, but on a neutron star, the reduction is quite large.

The empirical evidence in favor of General Relativity includes the experiments with clocks and the observations on the propagation of light that we have discussed in the preceding sections. It also includes the perihelion precession observed in the motion of planets, especially Mercury. This perihelion precession is a minor defect in the motion of planets which is in conflict with the Newtonian theory of motion. As is well known, according to Newtonian physics, a body orbiting in an inverse-square field of force has an elliptical orbit. In the absence of additional disturbances, the major axis of the ellipse would remain in a fixed orientation. But in the Solar System, the mutual forces among the planets disturb the elliptical orbits, and all the planetary orbits suffer some precession, that is, the major axes of the orbital ellipses gradually swing around the Sun, and correspondingly, the perihelion points also gradually swing, or precess, around the Sun (see Fig. 7.16). The perihelion precessions calculated from the mutual forces among the planets agree with the observed perihelion precessions except for a small deviation of 43 arcseconds per century in the case of Mercury (and smaller deviations for Venus, Earth, and Mars). This deviation was discovered by astronomers in the 19th century, and it remained a puzzle until Einstein recalculated the planetary motions in his theory of General Relativity and found that the combined effects of gravitational time dilation and the curvature of space give rise to an extra perihelion precession of the planetary ellipses, which accounts for the observed deviation.

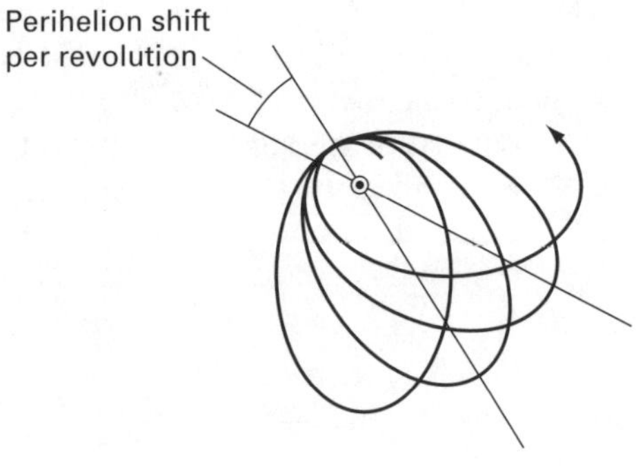

Fig. 7.16 Precession of an elliptical orbit around the Sun. The precession is exaggerated for the sake of clarity.

7.5 BLACK HOLES

Within the Solar System the effects of General Relativity are minuscule. We can readily understand this from the Schwarzschild solution, in which the general relativistic effects are contained in the factor $1 - 2GM/rc^2$. The relativistic effects are small as long as $2GM/rc^2$ is small compared with 1. Even on the surface of the Sun, the value of $2GM/rc^2$ is only about 10^{-6}. Large relativistic effects are found in the neighborhood of stellar objects more compact than the Sun, such as neutron stars. But the most spectacular relativistic effects are displayed by **black holes**, which are the remains of stars that have completely collapsed under their own weight, crushing their material to a singular state of infinite density. After such a collapse, nothing is left of the star except an extremely intense gravitational field, with extremely large relativistic effects.

The completely collapsed star is called a black hole because it is black; neither light nor anything else can escape the grip of the intense gravitational field. Particles, atoms, spacecraft, or astronauts can enter the black hole, but once they do so, they are trapped and can never emerge again. The intense gravitational field pulls them into the singularity, and they disappear.

From the Schwarzschild solution, we can understand how even light, which moves faster then anything else, is trapped by the black hole. According to Eq. (19), the speed of light is zero at $r = 2GM/rc^2$, a distance called the **Schwarzschild radius**. The zero speed of light results directly from the infinite time dilation at this radius. A light signal from the interior cannot proceed outward, and a light signal from the exterior cannot proceed inward--it stops and hovers at the Schwarzschild radius forever.

The boundary of the region of no escape that surrounds the black hole on all sides, permitting entrance but preventing exit, is called the **one-way membrane**, or the **horizon**. This membrane is not made of anything; it is not a surface of material but merely a place, the place of no return. A black hole of mass equal to that of the Sun has a one-way membrane with a Schwarzschild radius of 3 km. A black hole of mass equal to that of the Earth would have a one-way membrane with a Schwarzschild radius of just 9 mm (see Fig. 7.17)

The curvature of spacetime near the one-way membrane is large (but finite), and it becomes even larger inside the membrane. At the "center" of the black hole, spacetime develops a singularity, with infinite curvature. However, this "center" of the black hole is not a point

Fig. 7.17 A black hole of a mass equal to that of the Earth would have the size of this dot.

in space but a point in time. In the inside of the black hole, space and time are interchanged; the "distance" from the one-way membrane to the center is not a distance in space but a distance in time (the usual radial coordinate acts as a time coordinate, and the usual time coordinate acts as a space coordinate). Correspondingly, inside the black hole the spacetime geometry is not static; it is time-dependent, and it evolves, with a curvature which becomes larger and larger, until finally the curvature become infinite everywhere inside the black hole, and the spacetime geometry ceases to exist. This catastrophic time evolution inside the black hole proceeds very quickly. For instance, in a black hole of mass equal to that of the Sun, this time evolution takes only 10^{-5} s. Strangely, seen from the outside, the black hole does not evolve at all; its spacetime geometry is static, and it stays the same forever.

This mismatch between a violently evolving geometry in the interior of the black hole and an unchanging geometry in the exterior is camouflaged by the time dilation. As we saw in Section 7.2, a clock placed near a gravitating body runs slow, which represents one aspect of the distortion of spacetime. Near the Earth or the Sun the distortion of spacetime is fairly mild, and the time dilation is small. But near a black hole the distortion of spacetime and the time dilation are very large. At the one-way membrane the time dilation becomes infinite--a clock placed next to the membrane will run so slow that, compared with a clock far away from the black hole, it will appear to have stopped. If we imagine an astronaut in a spacecraft placed at the one-way membrane, all her life processes will be almost at a standstill--to the observers at mission control stationed far away from the black hole, she will appear to have frozen. Yet the astronaut will not perceive herself as having slowed down; instead, she will perceive the surrounding world as having speeded up.

Likewise, all the processes inside the black hole are subject to an infinite time dilation with respect to clocks outside the black hole. As judged from outside, any changes at or inside the one-way membrane take an infinite amount of time. Hence, as measured by outside observers, these changes never happen, and the black hole seems perfectly static.

Astronomers have found several objects in the sky that seem to be black holes. Among these is LMC X-1, a strong source of X rays in the Large Magellanic Cloud galaxy (see Figs. 7.18). LMC X-1 is a binary system consisting of an ordinary star and a very small compact object orbiting about each other with a period of 4.2 days. The ordinary star has a mass larger than 2 solar masses, and the compact object has a mass larger than 4 solar masses. This compact object is the source of the observed intense flux of X rays.

The compact object is believed to be a black hole. Of course, a black hole by itself cannot emit X rays (or anything else). But a black hole in a tight orbit around another star will produce a copious flux of

X rays by an accretion mechanism. The black hole will raise a large tidal bulge on the surface of the star (see Fig. 7.19), and it will pull a steady stream of gas off that surface. The gas spirals down toward the black hole and heats by friction against itself to temperatures of 10^8 or 10^9 K, high enough to emit X rays. The X rays that we observe are

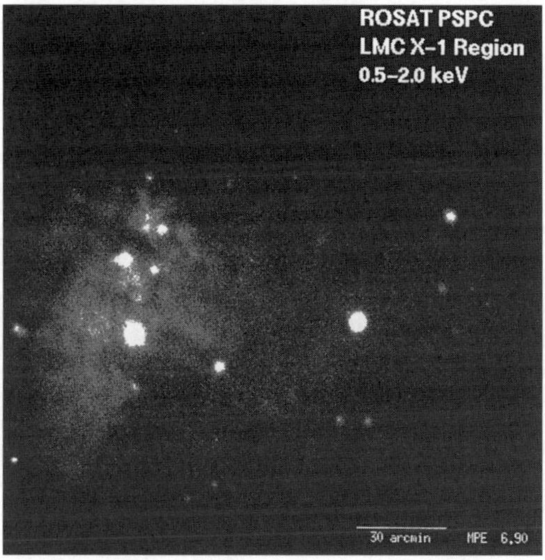

Fig. 7.18 X-ray image of LMC X-1 obtained by the ROSAT satellite. The bright source LMC X-1 (left of center) is surrounded by a glowing cloud of helium gas. (Max-Planck Institut für extraterrestrische Physik)

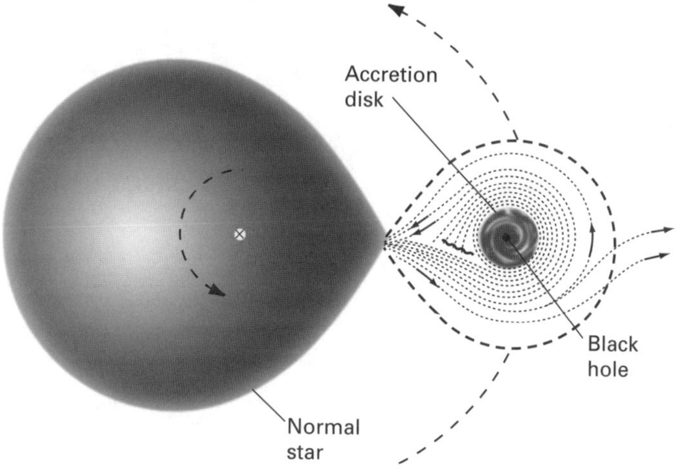

Fig. 7.19 The black hole orbiting the normal star in a binary system raises a pointed tidal bulge on the surface of the star. Gas spills from this tidal bulge and spirals toward the black hole, forming an accretion disk. (After Eardley and Press, [9])

emitted by the gas before it crosses the one-way membrane; any X rays emitted after the crossing will of course remain trapped in the black hole.

The evidence supporting the view that the compact object is a black hole is provided by detailed analysis of the X-ray flux. The characteristics of the X rays indicate that the source must be very small (by astronomical standards), no more than a few hundred kilometers across. But there is no way of confining a mass larger than 4 solar masses to such a small volume and keeping it in equilibrium. A mass with such a high density would immediately collapse under its own weight and form a black hole. The object in LMC X-1 must have collapsed long ago, and it is now a black hole.

Astronomers have found several other objects similar to LMC X-1 in the sky. All are members of binary systems, and they emit X rays by an accretion process similar to that of LMC X-1. There may also be many more black holes in isolation, without companion stars. But these would not emit a glow of X rays, and they would be undetectable except by their gravitational effects. Isolated black holes are the hidden reefs of outer space. They can be spotted only when they act as gravitational lenses, deflecting rays of light that pass nearby.

REFERENCES

1. A. Pais, *'Subtle is the Lord . . .' The Science and the Life of Albert Einstein* (Oxford University Press, Oxford, 1982), p. 179

2. H. C. Ohanian and R. Ruffini, *Gravitation and Spacetime* (Norton, New York, 1994), Chapter 4.

3. R. V. Pound and G. A. Rebka, Phys. Rev. Lett. **4**, 337 (1960).

4. R. V. Pound and J. L. Snider, Phys. Rev. Lett. **13**, 539 (1964); Phys. Rev. B **140**, 788 (1965).

5. R. F. C. Vessot, M. W. Levine, E. M. Mattison, E. L. Blomberg, T. E. Hoffman, G. U. Nystrom, B. F. Farrel, R. Decher, P. B. Eby, C. R. Baugher, J. W. Watts, D. L. Teuber, and F. D. Wills, Phys. Rev. Lett. **45**, 2081 (1980).

6. J. C. Häfele and R. E. Keating, Science **177**, 168 (1972).

7. C. O. Alley, "Relativity and Clocks," in *Proceedings of the 33rd Annual Symposium on Frequency Control* (Electronic Industries Association, Washington, D.C., 1979).

8. I. I. Shapiro et al., Phys. Rev. Lett. **20**,1265 (1968); Phys. Rev. Lett. **26**, 1132 (1971).

9. D. M. Eardley and W. H. Press, Ann. Rev. Astron. Astrophys. **13**, 381 (1975).

PROBLEMS

1. The Earth is in free fall in the gravitational field of the Sun, the Moon, and the other planets. Does this completely eliminate all the external gravitational effects in

the freely falling reference frame of the Earth? (*Hint*: What causes the tides of the oceans?)

2. According to a rumor, there are some apple trees in Tibet that produce apples of negative mass.* How does the attachment of such an apple to a branch differ from that of an ordinary apple? Draw a picture of the apple, its stem, and the branch to which it is attached. In which direction does such apple accelerate if it becomes detached from its branch?

3. Consider a charged particle held at rest in the gravitational field of the Earth. Relative to a freely falling local inertial reference frame, such a particle has an acceleration. According to the Principle of Equivalence, the particle should then radiate electromagnetic radiation. However, relative to the reference frame centered on the Earth, the particle has no acceleration and hence cannot radiate. Try to resolve this paradox.

4. Calculate the fractional blueshift for a light or radio signal emitted at a point 10,000 km above the surface of the Earth and received on the surface. Note that for such a height, the expression for the Earth's gravitational potential is $\Phi = -GM/r$.

5. According to the Principle of Equivalence, the fictitious centrifugal force felt in a rotating reference frame has the same effect as a centrifugal gravitational force. The gravitational potential that corresponds to this centrifugal gravitational force is $\Phi = -\frac{1}{2}\omega^2 r^2$, where ω is the angular velocity of rotation and r is the distance from the axis of rotation.
(a) Verify that for this potential, the gravitational force $-m\, d\Phi/dr$ on a particle of mass m agrees with the familiar centrifugal force $m\omega^2 r$.
(b) According to Eq. (4), what is the fractional frequency shift for a light signal emitted at $r = r_1$ and received at $r = r_2 < r_1$?
(c) Show that this frequency shift agrees with the frequency shift predicted by Special Relativity for clocks with the rotational speeds ωr_1 and ωr_2. This means we have two independent, but consistent, ways to calculate this frequency shift: We can calculate it in a rotating, noninertial reference frame (in which the clocks are at rest but experience a "gravitational" force); or we can calculate it in a nonrotating, inertial reference frame (in which the clocks are in motion, and there is no gravitational potential).

6. In Section 3.5 we resolved the twin paradox of Special Relativity by arguing that in the reference frame of Stella (the traveling twin), at the instant that Stella decelerates and reverses her motion, the clocks of Terra (the twin on Earth) jump ahead, making her older. This can be explained in more detail as a gravitational time dilation, since in the reference frame of Stella, during the deceleration there is a fictitious inertial force, and according to the Principle of Equivalence, this has the

* I am indebted to R. Lichtenstein for this amusing tale.

same effect as a uniform gravitational field extending over all of space. In this effective gravitational field in the decelerating reference frame of Stella, which twin is at the higher gravitational potential? Whose clock runs faster?

7. Because of the rotational motion of the Earth, a clock placed on the equator has a speed of about 460 m/s relative to a clock placed on the pole. Because of the equatorial bulge of the Earth, the clock at the equator is about 21 km farther from the center of the Earth than is the clock at the pole. Calculate the difference between the time dilation factors for these two clocks. Take into account both the special-relativistic time dilation due to the velocity of the equatorial clock in the inertial reference frame centered on the Earth and the gravitational time dilation. Pretend that over the height difference of 21 km, the value of g is constant. (Note that the two time-dilation effects almost cancel--the two clocks keep nearly the same time. A more exact calculation, taking into account the shape of the equatorial bulge and its gravitational potential, shows that clocks placed at sea level at the equator and the pole keep the same time, and so do all clocks placed at sea level anywhere on Earth.)

8. Consider a clock that is moving with some speed v in the gravitational field of a spherical mass M. According to Eqs. (11) - (13), the proper time registered by the clock is

$$d\tau = \sqrt{1 - \frac{2GM}{rc^2} - \left(1 + \frac{2GM}{rc^2}\right)\frac{v^2}{c^2}}\; dt$$

(a) Show that this is approximately

$$d\tau \simeq \left(1 - \frac{GM}{rc^2} - \frac{1}{2}\frac{v^2}{c^2}\right)dt$$

(b) The term GM/rc^2 represents the time dilation due to gravitation, and the term $\frac{1}{2}v^2/c^2$ represents the time dilation due to the speed. Consider a clock placed on the equator of the Earth, with a rotational speed of about 460 m/s. Which term is larger? What is the sum of the two terms?

(c) Show that for a clock on a satellite in a circular orbit around a mass M,

$$d\tau \simeq \left(1 - \frac{3}{2}\frac{GM}{rc^2}\right)dt$$

What must be the radius of the orbit of an artificial Earth satellite if its time dilation is to match the time dilation of a clock placed on the equator?

9. According to Eqs. (11) and (13), the infinitesimal length increment dl that corresponds to an infinitesimal increment dx along the positive x axis is

$$dl = \sqrt{-g_{11}}\,dx = \sqrt{1 + \frac{2GM}{rc^2}}\,dx \simeq \left(1 + \frac{GM}{rc^2}\right)dx = \left(1 + \frac{GM}{xc^2}\right)dx$$

(a) Consider a finite increment in x, from $x = r_1$ to $x = r_2$. Evaluate $\int dl$ to find the actual distance between the points r_1 and r_2.
(b) What is the distance between the points $r_1 = 7000$ km and $r_2 = 12{,}000$ km near the surface of the Earth? How much does it differ from 5000 km?
(c) What is the actual distance between the points $r_1 = 7 \times 10^5$ km and $r_2 = 12 \times 10^5$ km near the surface of the Sun? How much does it differ from 5×10^5 km?

10. (a) Show that according to Eqs. (11) and (13), the actual volume of a small cube with increments dx, dy, dz as its sides is

$$dV = \left(1 + \frac{2GM}{rc^2}\right)^{3/2} dx\,dy\,dz \simeq \left(1 + \frac{3GM}{rc^2}\right) dx\,dy\,dz$$

(b) Consider the atmosphere of the Earth, which has a volume of approximately (area of Earth) $\times$ (height) $\simeq 4\pi R^2 \times 100$ km, where $R \simeq 6400$ km is the Earth radius. Use the result of part (a) to find the discrepancy between the actual volume of the atmosphere and the volume calculated according to Euclidean geometry. For this problem, it is an adequate approximation to treat the radial coordinate as constant, $r \simeq 6400$ km, over the entire height of the atmosphere.

11. In Newtonian mechanics we define the **escape velocity** for a projectile in the gravitational field of a planet or a star as the least velocity with which the projectile must be launched if it is to escape from the gravitational field to infinity (a parabolic orbit).
(a) Show that the Newtonian escape velocity from an initial position r is

$$v = \sqrt{2GM/r}$$

(b) If we assume that the maximum attainable speed is c, what is the smallest radius from which escape is possible? Compare with the Schwarzschild radius.

12. At a radius $r = 3GM/c^2$ around a black hole, a light signal is deflected so much that it orbits the black hole. According to Newtonian mechanics, what is the period of this orbit? (By coincidence, the Newtonian result is actually correct.)

13. According to astronomical observations, the X-ray source Cygnus X-1 is a binary star system one of whose members has a mass of about 10 times the mass of the Sun and a radius of at most a few hundred kilometers. What can you conclude about this star?

14. The masses of the two bodies in the binary X-ray source Cygnus X-1 are about 30 times the mass of the Sun and 10 times the mass of the Sun. The period of the motion is 5.6 days. Calculate the radial distance between these bodies. Assume that both bodies are in circular orbits about the common center of mass.

Appendix I: Textbooks and Further Reading

The first two parts of this appendix recommend some other textbooks and some biographies of Einstein. The third part lists further readings to supplement individual chapters. With some exceptions, the emphasis is on recent publications. For a complete review of older publications, see "Resource Letter SRT 1 on Special Relativity Theory" by G. Holton, Am. J. Phys. **30**, 462 (1962).

1. RECOMMENDED TEXTBOOKS

The following two books provide excellent alternative introductions to relativity at the same level as the present book:

E. F. Taylor and J. A. Wheeler, *Spacetime Physics* (Freeman, New York, 1964 and 1992)--This book was the first to emphasize that relativity is a theory of the geometry of spacetime. Wheeler's favorite method of explanation is to pose a puzzle or an apparent paradox and then weave the explanation into the resolution of this puzzle. The book is written in a charmingly informal style, and it is enlivened with many short dialogues that try to anticipate questions the readers might ask.

A. P. French, *Special Relativity* (Norton, New York, 1968)--A very clear, straightforward, and well-organized introduction. Includes an excellent and careful treatment of the implications of relativity for electricity and magnetism. However, the discussion of the experimental evidence for various aspects of relativity is outdated.

Some other books at about the same level are

W. Rindler, *Essential Relativity* (Springer-Verlag, New York, 1977)--The first part of the book deals with Special Relativity, presenting a clear and incisive critical examination of the conceptual foundations of the theory. It contains a concise treatment of four vectors and their application to relativistic mechanics.

G. Barton, *Introduction to the Relativity Principle* (Wiley, New York, 1999)--Provides complete and competent coverage of Special Relativity, including some of the modern experimental evidence. Excellent discussion of the differences between Newtonian and relativistic physics in kinematics and dynamics and thorough treatment of collisions. But the book does not convey the essential geometrical character of the theory, and spacetime diagrams are given short shrift.

G. F. R. Ellis and R. M. Williams, *Flat and Curved Space-Times* (Clarendon Press, Oxford, 1988)--A simple and clear introduction to the geometry of spacetime with an abundance of spacetime diagrams. Approaches relativity via the study of Doppler shifts ("K calculus") and extracts the time dilation and length contraction from this, with the Lorentz transformation derived only as an afterthought.

L. Sartori, *Understanding Relativity* (University of California Press, Berkeley and Los Angeles, 1996)--Good explanations of basic concepts and careful examination of several paradoxes. Uses Loedel diagrams instead of Minkowski diagrams for the graphical representation of Lorentz transformations. In the Loedel diagram both pairs of axes x,ct and x',ct' are slanted symmetrically (see Problem 2.16), but this has the serious disadvantage that an alteration of the velocity V of the transformation requires rearrangement of the ct axis and redrawing of all the worldlines.

More advanced treatments of relativity can be found in the following books, most of which also include general relativity:

P. Bergmann, *Introduction to the Theory of Relativity* (Prentice-Hall, Englewood Cliffs, NJ, 1942)--Contains a thorough introduction to the general tensor calculus. This calculus permits a concise formulation of the equations of special relativity, and it is essential for general relativity.

W. L. Burke, *Spacetime, Geometry, and Cosmology* (University Science Books, Mill Valley, CA, 1980)--Chapter 2 of this book provides a simple and clear introduction to the calculus of differential forms, which is an elegant alternative to tensor calculus.

W. Pauli, *Theory of Relativity* (Pergamon Press, London, 1958)--A classic written when Pauli was still a student. It originally was published in the *Mathematical Encyclopedia* as a complete review of all the advances in relativity up to 1921, and although some parts of it are now outdated, most of the theoretical material remains valid and informative.

C. Møller, *The Theory of Relativity* (Clarendon Press, Oxford, 1952)--Very thorough treatment, including detailed discussions of relativistic energy, momentum, angular momentum, and center of mass for arbitrary systems.

J. L. Synge, *Relativity: The Special Theory* (North-Holland, Amsterdam, 1965)--Good discussion of general systems of particles and collisions, and also gives a simple and clear definition of the energy-momentum tensor.

2. BIOGRAPHIES

A. Pais, *Subtle Is the Lord...* (Oxford University Press, Oxford, 1982)--A scientific biography that focuses on Einstein's contributions to physics and on the interplay between Einstein's work and that of his contemporaries. Although the writing lacks polish, the book is very informative.

B. Hoffmann, *Albert Einstein: Creator and Rebel* (Viking Press, New York, 1972)--A brilliantly written biography of Einstein.

A. P. French, ed., *Einstein: A Centenary Volume* (Harvard University Press, Cambridge, MA, 1979)--Contains a collection of reminiscences about Einstein by physicists who worked with him.

H. Woolf, ed., *Some Strangeness in the Proportion* (Addison-Wesley, Reading, MA, 1980)--The last chapters offer some more reminiscences about Einstein.

3. FURTHER READINGS

Chapter 1. Relativity in Newtonian Mechanics and the Michelson-Morley Experiment

The story of the 18th-century attempts to determine geographical longitude by synchronized clocks and the construction of suitable portable clocks by the English clock maker Harrison is told by D. Sobel, *Longitude* (Walker and Company, New York, 1995). Sobel writes in a charming and lively manner, but she fails to include relevant technical details about Harrison's clocks. Some of these details can be found in D. Sobel and W. J. H. Andrewes, *The Illustrated Longitude* (Walker and Company, New York, 1998), an expanded version of the book.

The crucial role of Newton's Second Law in the synchronization of clocks in inertial reference frames seems not to have been widely recognized. But this law enters implicitly into the mechanical synchronization methods proposed by J. Aharoni, *The Special Theory of Relativity* (Oxford University Press, Oxford, 1965) and by Ya. P. Terletskii, *Paradoxes in the Theory of Relativity* (Plenum Press, New York, 1968). Aharoni proposes synchronization by means of "bullet signals" fired from standard guns, which ensure equal speeds of the bullets. Terletskii proposes a method based on the equal speeds attained by two particles of equal mass exerting a repulsive force on each other. Both methods are impractical, but in principle they are consistent with synchronization by clock transport.

Fundamental aspects of space and time are examined by H. Reichenbach in *The Philosophy of Space and Time* (Dover Publications, New York, 1957). This book gave rise to a lengthy controversy regarding clock synchronization, because Reichenbach argued that the alteration of clock synchronization by an arbitrary amount linearly proportional to distance (in some given direction) is consistent with Newton's First Law and that, therefore, we cannot fix the clock synchronization uniquely except by adopting some convention. Many philosophers of science (and even a few physicists) accepted this mistaken "conventionalist thesis"; for a review and a nearly exhaustive collection of references, see R. Anderson, I. Vetharaniam, and G. E. Stedman, Phys. Reports **295** 93 (1998). Although it is indeed true that Newton's First Law is consistent with a linear alteration of synchronization, Newton's Second Law is *not* consistent with any such alteration, and it therefore uniquely fixes the synchronization (as discussed in Section 1.1 of this book, Newton's Second Law fixes the synchronization to coincide with clock transport).

A brief survey of 19th-century wave theory and experiments to detect the motion of the Earth relative to the ether can be found in

Møller, op. cit., Chapter 1.

An encyclopedic survey of the development of the wave theory of light and the 19th-century attempts at mechanical models of the ether are presented by E. Whittaker, *A History of the Theories of Aether and Electricity* (Nelson, London, 1951). Volume I of this history is competent and reliable, but Volume II makes a silly attempt to assign all of the credit for the discovery of the theory of relativity to Lorentz and Poincaré and is not to be trusted. For critical commentary on this point, see G. Holton, Am. J. Phys. **28**, 627 (1960).

The 19th-century speculations regarding the ether are discussed in the article "Aether" by J. Larmor in the *Encyclopaedia Britannica* (11th edition).

A good historical account of the Michelson-Morley experiment is given by Pais, op. cit., Chapter 6. Interesting details about Michelson's initial experiment and his activities at Berlin and Potsdam can be found in R. S. Shankland, Am. J. Phys. **32**, 16 (1964).

A valuable summary of the Michelson-Morley experiments performed up to 1930 is given by R. S. Shankland, S. W. McCuskey, F. C. Leone, and G. Kuerti, Rev. Mod. Phys. **27**, 167 (1955). Summaries of later versions of the Michelson-Morley experiment and other experiments testing Special Relativity are given by D. Newman, G. W. Ford, A. Rich, and E. Sweetman, Phys. Rev. Lett. **40**, 1355 (1878), and by D. W. MacArthur, Phys. Rev. A **33**, 1 (1986). For details on the experiments, it is necessary to consult the original articles, which are listed in the References at the end of Chapters 1, 2, and 3.

Chapter 2. Einstein's Theory and the Lorentz Transformation

The historical background for Einstein's formulation of Special Relativity is described by S. Bergia in A. P. French, ed., op. cit., and by Pais, op. cit. Both of these references summarize and evaluate the contributions of Lorentz and of Poincaré. Pais attributes the first derivation of the Lorentz transformation equations to Voigt, but this is a mistake, because Voigt's transformation equations did not include the factor γ. Pais emphasizes the important role played by Planck, who became the first convert to relativity and whose influence as the preeminent German professor was crucial in bringing relativity to the attention of other physicists.

Einstein's own account of how he came upon relativity is contained in his "Autobiographical Notes" in P. A. Schilpp, ed., *Albert Einstein: Philosopher-Scientist* (Harper Brothers, New York, 1959).

Einstein's article "Zur Elektrodynamik bewegter Körper" in Ann. Phys. **17**, 891 (1905) is well worth reading, especially the first part. This is his original article on relativity, and it is amazing for its completeness: it includes the definition of simultaneity, the Lorentz transformations, the length contraction and the time dilation, the velocity combination law, the Doppler-shift formula, the transformations of electric and magnetic fields, and the expression for the relativistic kinetic energy. However, the article commits a mistake in the expression for the "transverse" mass. A translation of Einstein's article is available in H. A. Lorentz, A. Einstein, H. Minkowski, and H. Weyl, *The Principle of Relativity* (Methuen, London, 1923), which is a collection of reprints of original papers. Oddly, this collection does not include the work of Planck, who was Einstein's first and most influential follower.

There are many alternative methods for deriving the Lorentz transformation equations. Perhaps the most widely used method is to require that the transformation from x,t to x',t' be linear and that the speed of light be invariant, that is, $x = ct$ and also $x' = ct'$ for a light signal emitted from the origin. This is the method adopted by French, op. cit., and, with minor variations, by many other authors.

Another method is to use the invariance of the speed of light to show that s^2 is invariant and then derive the Lorentz transformation equations from the requirement of invariance of s^2. This is the method favored by Taylor and Wheeler, op. cit.

A clever graphical method (different from that in Section 2.3) is given by T. Schwartz, *Relativity in Illustrations* (Dover, New York, 1989).

There are several other interesting derivations that take as their starting point not the invariance of the speed of light but the existence of an invariant upper limit to the particle velocity. For instance, see Terletskii, op. cit., pp. 17-26; Rindler, op. cit., pp. 51-53; A. R. Lee and T. M. Kalotas, Am. J. Phys. **43**, 434 (1975); and J.-M. Lévy-Leblond, Am. J. Phys. **44**, 271 (1976).

A delightful discussion of spacetime geometry, with emphasis on concepts rather than mathematics, is given by J. L. Synge, *Talking About Relativity* (North-Holland, Amsterdam, 1970).

Chapter 3. Time Dilation and Length Contraction

A wide-ranging discussion of the twin paradox, including arguments and counterarguments from opponents of relativity, can be found in L. Marder, *Time and the Space Traveller* (University of Pennsylvania Press, 1974)

R. C. Tolman, *Relativity, Thermodynamics, and Cosmology* (Clarendon Press, Oxford, 1934), p. 194, shows how to resolve the twin paradox by using the Principle of Equivalence to calculate clock rates in the accelerated reference system of the traveling twin.

A generous selection of examples of the distorted visual appearance of fast-moving bodies in photographs is contained in D. E. Mook and T. Vargish, *Inside Relativity* (Princeton University Press, 1987). References to many other articles with further examples are listed in U. Kraus, Am. J. Phys. **68**, 1 (2000). Most of these articles give the misleading impression that the distortions are a relativistic effect, whereas in fact, as mentioned in Section 3.4, the visual distortions arise largely from time delays, which would be present even in the Newtonian case.

A variety of puzzles and paradoxes arising from time dilation, length contraction, and other aspects of relativity are discussed in Ya. P. Terletskii, op. cit. The problem sections in Taylor and Wheeler, op. cit., contain a few more paradoxes.

Chapter 4. Relativistic Mechanics

In the last section of his 1905 paper, Einstein made an attempt at finding the relativistic expression for momentum, in the guise of "transverse" and "longitudinal" mass. He obtained the correct longitudinal mass but the wrong transverse mass. His mistake was quickly corrected by Planck, who obtained the correct expression (4.8) for the relativistic momentum. Brief summaries of Planck's derivation can be found in Pais, op. cit., p. 150, and in Pauli, op. cit., p. 82.

The collision argument used in Section 4.1 to derive the formula for the relativistic momentum is due to R. N. Lewis and C. Tolman, Phil. Mag. **18**, 510 (1909). This argument is used in many other textbooks, with minor variations. An elegant version of the argument, with the proper time as parameter, is presented by Taylor and Wheeler, op. cit., first edition.

An alternative derivation relies on the four-vector property of the conservation law of energy and momentum; for instance, see Taylor and Wheeler, op. cit., second edition, and Barton, op. cit.

The relation between energy and mass was originally obtained by Einstein by a somewhat different argument involving the emission of two opposite pulses of light by a body at rest. This argument is carefully explained in Sartori, op. cit. Einstein later contrived several other arguments, including the argument we used in Section 4.4.

The origins of the relationship between inertia and energy are nicely traced by M. von Laue in the essay "Inertia and Energy" in Schilpp, ed., op. cit.

A historical and logical critique of mass and inertia in relativity is given by M. Jammer, *Concepts of Mass in Classical and Quantum Physics* (Harvard University Press, Cambridge, MA, 1961), Chapters 12 and 13. Jammer lists all of the early experimental measurements of variation of mass with speed. These experiments attempted to discriminate between the predictions of relativity theory and the (older) theory of Abraham, which sought to construct the electron mass from its electrostatic self-energy. In their review of these experiments, P. S. Faragó and L. Jánossy, Nuovo Cimento **5**, 1411 (1957) conclude, "...Regarding the fairly large number of direct experiments on the behaviour of free electrons we could hardly find such results which would prove the validity of the relativistic relation with a margin of error much less than, say, the difference between the results of relativity theory and the theory of Abraham..." They also say that the only high-precision confirmation of the relativistic law of the variation of electron mass with speed is found in measurements of the fine-structure splitting in the spectra of hydrogenlike atoms. The latter statement hinges on Sommerfeld's relativistic version of the Bohr model, which attributes the fine structure to the relativistic mass. However, the Bohr model is not applicable to the low-lying states of the hydrogen atom. Instead, we must use the Dirac equation, according to which the fine structure is attributed not to the relativistic mass but to the spin-orbit coupling of the electron. Thus the fine-structure splitting confirms the Dirac equation and the spin of the electron, but it does not directly confirm the formula for the relativistic mass.

Chapter 5. Collisions and Conservation of Energy-Momentum

Four-vector techniques for the analysis of elastic and inelastic collision are carefully presented at an introductory level in Taylor and Wheeler, op. cit., and at a somewhat more advanced level in Barton, op. cit., and in R. Hagedorn, *Relativistic Kinematics* (Benjamin, New York, 1964).

Relativistic rocket motion and implications for space travel are discussed by S. von Hoerner, "The General Limits of Space Travel," in A. G. W. Cameron, ed., *Interstellar Communication* (Benjamin, New York, 1963).

Applications of relativity to engineering are summarized by W. K. H. Panofsky, "Special Relativity Theory in Engineering," in H. Woolf, op. cit.

Chapter 6. Electromagnetism and Relativity

The following articles show several ways to obtain the Maxwell equations by starting with Coulomb's Law and performing Lorentz transformations on the electric and magnetic fields:

D. H. Frisch and L. Wilets, Am. J. Phys. **24**, 574 (1956)
R. S. Elliot, IEEE Spectrum, March 1966, p. 140
J. R. Tessman, Am. J. Phys. **34** (1966)
D. H. Kobe, Am. J. Phys. **54**, 631 (1986)

Our determination of the magnetic field of a uniformly moving charge in Section 6.3 imitates the article by Frisch and Wilets, but our determination of the field of an accelerated charge in Section 6.5 is simpler (and less rigorous) than that of Frisch and Wilets.

The goal of obtaining Maxwell's equations can also be achieved by performing Lorentz transformations on the potentials instead of on the fields. This leads to some simplifications of the calculations, as shown in the following books:

M. Schwartz, *Principles of Electrodynamics* (McGraw-Hill, New York, 1972), Chapter 3.
H. C. Ohanian, *Classical Electrodynamics* (Allyn and Bacon, Newton, MA, 1988), Chapter 8.

Chapter 7. General Relativity

Excellent qualtitative discussions of General Relativity and curved spacetime can be found in

P. Bergmann, *The Riddle of Gravitation* (Charles Scribner's Sons, New York, 1968)
R. Geroch, *General Relativity from A to B* (University of Chicago Press, Chicago, 1978)
G. F. R. Ellis and R. M. Williams, *Flat and Curved Space-Times* (Oxford University Press, Oxford, 1988)

A nice elementary introduction to General Relativity is given by A. P. French, "The story of general relativity," in A. P. French, ed., op. cit. This includes an interesting historical account of the first expeditions to measure the light deflections during eclipses.

A fascinating explanation of the geometrical meaning of Einstein's gravitational equation, without the mathematical apparatus of tensor calculus, is given by J. A. W. Wheeler, *A Journey Into Gravity and Spacetime* (Scientific American Library, New York, 1990). The first part of the book provides a clear and amusing introduction to spacetime, worldlines, and geodesics. Wheeler then reformulates the Einstein equation so that it resembles Gauss' Law, and in a remarkable display of virtuosity, he obtains the exact Schwarzschild solution without using tensor calculus or partial differential equations.

At a somewhat more advanced level, good introductions are M. Berry, *Principles of Cosmology and Gravitation* (Cambridge University Press, Cambridge, 1976) and I. R. Kenyon, *General Relativity* (Oxford University Press, Oxford, 1990). Kenyon succeeds in presenting Einstein's equation with only a minimal amount of tensor calculus. Both Kenyon and Berry obtain the Schwarzschild solution by (more or less) inspired guesswork rather than by the actual solution of the Einstein equation.

Appendix II: Answers to Even-Numbered Problems

Chapter 1

8. 4.0×10^8 m/s; no

10. $v' = 26.3$ mi/h; $\theta = 32.5^0$ north of east

12. $\Delta K' = \frac{1}{2}mv_2{}^2 - \frac{1}{2}mv_1{}^2 + mv_1V - mv_2V$;

$F \times (x'_2 - x'_1) = F \times (x_2 - x_1) + mv_1V - mv_2V$; yes

14. $K' = \frac{3}{2}NkT + \frac{1}{2}mV^2$; $K' = \frac{1}{2}mV^2$

16. (b) 10^{-4} radians or 21 arcsec

18. $\Delta\phi \simeq \frac{2\pi L}{\lambda}\frac{V^2}{c^2}\cos 2\theta$

20. (a) $\frac{2\pi c}{\lambda}(t_2 - t_1) = \frac{2\pi L}{\lambda}\left[n^2\frac{V}{c} + n - 1 - \frac{V}{c}\right]$

(b) $\frac{2\pi c}{\lambda}(t_2 - t_1) = \frac{2\pi L}{\lambda}\left[n - 1 - \frac{V}{c}\right]$

(c) $\frac{2\pi c}{\lambda}(t_2 - t_1) = \frac{2\pi L}{\lambda}\left[n - 1 + \frac{V^2}{c^2}\frac{1}{n} - \frac{V^2}{c^2}\right]$

Chapter 2

2. (a) Angle between the x,x' axes is $\theta = 17^0$; (b) $x' \simeq 0.84$, $ct' \simeq 3.6$; (c) $v'_x = 0.23c$; (d) $x' = 0.839$, $t' = 3.56$, $v'_x = 0.235c$

4. Tucana is first, by 6.1 hours

6. $V/c > 0.51$

8. (a) $-\omega R\Delta x/c^2$, $-2\omega R\Delta x/c^2$, $-(n - 1)\omega R\Delta x/c^2 \simeq -2.1 \times 10^{-7}$ s; (b) $+\omega R\Delta x/c^2$

10. $\Delta\phi' \simeq -\frac{VL}{c^2}\omega$

20. No; yes; time interval is 4.4 s

26. $v_0 = \dfrac{3v_b + v_b{}^3/c^2}{1 + 3v_b{}^2/c^2}$

30. $\phi \simeq 10^{-4}$ radians or 21 arcsec

34. $t = 0.19$ year; $x = 9.3 \times 10^{15}$ m = 0.99 light-year

Chapter 3

2. $\sqrt{1 - V^2/c^2} \simeq 1 - 6.7 \times 10^{-10}$

4. 105 h

6. 3×10^{-8} s

8. (a) 2.2×10^{-27}; (b) 1.7×10^{-4}

10. (a) 26 years; (b) 24 years

r

ars; 0.56 light-year; $v_x = 0.77c$;

year; 1.5×10^4 light-years; $v_x = (1 - 2.2 \times 10^{-9})c$

$V^2/c^2/(c - V)$, $L\sqrt{1 - V^2/c^2}/(c + V)$,

$c\sqrt{1 - V^2/c^2}$; yes

20. $v/c = 0.92$

24. 0.147 m

26. - 0.064 m; $- 5.5 \times 10^{12}$ m^3

28. 0.901 m

30. 0.12 m, 0.20 m, 0.23 m; 59^0 and 31^0

32. (a) $(\sqrt{1 - V^2/c^2} + \sqrt{1 - 4V^2/c^2})L/V$

(b) $(1 - 2V^2/c^2 + \sqrt{1 - 5V^2/c^2 + 4V^4/c^4})L/V$

(c) $(1 - 2V^2/c^2 + \sqrt{1 - 5V^2/c^2 + 4V^4/c^4})L/V$; least in reference frame of each spaceship

34. No

36. The rod slides through the manhole at an angle (see Figure 3.24b).

38. The instantaneous configuration of the rod is bent along a downward parabolic arc, as illustrated in Fig. 3.25, and the rod "wriggles" out of the gap

Chapter 4

2. Nearly 90^0

4. $v/c = 0.758$, 0.875, 0.948, and 0.996

6. $\dfrac{3v^2}{4c^2} = 5.2 \times 10^{-7}$; $\dfrac{v^2}{2c^2} = 3.5 \times 10^{-7}$

10. (a) $K = 1.4 \times 10^{18}$ J, about 1.4% of yearly US energy expenditure; (b) $K = 1.4 \times 10^{13}$ J

12. (a) 4.7×10^{-4} kg; (b) no

14. $B^2/2\mu_0 = 4.0 \times 10^{11}$ J/m^3, 4.4×10^{-6} kg/m^3

16. 13.6 eV/c^2, 1.4×10^{-6} %

18. 8.5 MeV

20. The photons of the light wave have energy $E = h\nu$; hence the transformation law for E must be the same as that for ν

24. $p \simeq E/c = 1000$ GeV/c, 3.3 T

26. $p = 64$ GeV/c, $E = 64$ GeV, $v \simeq c$, 0.14 s

28. $\nu = 7.6 \times 10^{-4}$ Hz; $r = 6.3 \times 10^{10}$ m

30. 0.056 T and 0.91 T; 3.2×10^8 Hz and 3.2×10^8 Hz

Chapter 5

2. (c) 0.256 MeV

6. 5.4 keV

8. 19.9 MeV

12. $E = mc^2 = 0.511$ MeV; $p = E/c = 0.511$ MeV/c; westward

14. $Q = - 23.8$ MeV; no, Q is negative

16. $Q = -0.6$ MeV and $Q = 1.9$ MeV; ^{5}Li is unstable, and ^{6}Be is stable
18. $V = 0.75c$; $V = 0.48c$ in Newtonian physics
20. $Q = 135$ MeV, $E' = 67$ MeV each in rest frame; $E = 318$ MeV and $E = 14$ MeV in laboratory frame
22. $\Sigma \bar{m}_i c^2 = 194$ GeV; 40,000 GeV
26. 190 GeV; 0.96c
28. 0.767 GeV
30. (a) 1.784 GeV; (b) 12 TeV
34. $\Delta m = W/c^2 - \sqrt{W^2/c^4 + m^2} + m.$
36. (c) For 0.9c, $E_{rad} = 1.5 \times 10^{22}$ J, $K = 1.2 \times 10^{22}$ J;
for 0.99c, $E_{rad} = 5.9 \times 10^{22}$ J, $K = 5.5 \times 10^{22}$ J;
rocket propulsion requires an energy of fuel twice as large as E_{rad}

Chapter 6

2. $E = \dfrac{q}{4\pi\epsilon_0} \dfrac{\gamma r}{[\gamma^2 r^2 \cos^2\theta + r^2 \sin^2\theta]^{3/2}}$.

6. $E = \dfrac{1}{4\pi\epsilon_0} \dfrac{e}{r^2} = 5.3 \times 10^{11}$ V/m; $B' = 13$ T

8. (a) $E = \dfrac{\sigma}{2\pi\epsilon_0} \ln \left(\dfrac{r + d/2}{r - d/2}\right)$

(b) $B' = \mu_0 \dfrac{I}{2\pi d} \ln \left(\dfrac{r + d/2}{r - d/2}\right)$

10. No; yes

14. (a) $F = qv\mu_0 \dfrac{I}{2\pi r}$, $a = \dfrac{qv\mu_0}{m} \dfrac{I}{2\pi r}$;
(b) $E' = v\mu_0 I/2\pi r$, same force and acceleration as in (a);
(c) $\lambda' = \epsilon_0 \mu_0 v I$

Chapter 7

2. Apple "hangs" upward; if released from branch, apple falls downward
4. $(\nu_2 - \nu_1)/\nu_2 = 4.3 \times 10^{-10}$
6. Terra is at high gravitational potential, and her clock runs faster
8. (b) The gravitational term is much larger, and the sum is approximately 6.9×10^{-10}; (c) $r = 3R/2 = 9.6 \times 10^6$ m
10. (b) 110 km^3
12. $T = 6\pi\sqrt{3}\, GM/c^3$
14. $r = 3.2 \times 10^{10}$ m

Index